A Short History of Mechanics

Second Edition

A Short History of Mechanical Engineering

Second Edition

W. F. Greaves M.C.C. Ed.

Adviser in Technical Education
Mauritius Institute of Education
Formerly Chairman of Examiners in Engineering
Workshop Theory and Practice, Joint Matriculation Board

J. H. Carpenter B.A.

Lecturer in General Studies
Worcester Technical College

Longman

LONGMAN GROUP LIMITED
London

Associated companies, branches and representatives throughout the world

First published 1969
Second edition 1978
Second impression 1979

ISBN 0 582 35143 X

Printed in Hong Kong by
Commonwealth Printing Press Ltd

Foreword

The authors have endeavoured to provide, under one cover, a compact account of the history of mechanical engineering. In a work of this size a degree of selectivity has been unavoidable considering the scope and complexity of the subject. It is hoped, however, that the book will be of particular value to students in schools and technical colleges and as an introductory reference work in libraries.

The book covers the requirements of the appropriate sections of the Joint Matriculation Board's syllabuses in Metalwork and Engineering Workshop Theory and Practice and it contains a breadth of information which should prove useful to students taking examinations set by the other examining bodies. Much of the material in the work should be of help to social studies classes and to students taking economic history.

W.F.G.

J.H.C.

Foreword to the Second Edition

The increasing interest in the historical aspects of technology and the reception given to the first edition of this book has encouraged the preparation of a second edition. While no radical changes have been made in the format or the content, the opportunity to revise certain sections, particularly in the light of recent research, has been taken. Some of the illustrations have been replaced with better material and a number of new ones have been added.

Metric units or approximate metric equivalents have been inserted except in the cases where to do so would have made historical nonsense, e.g. Watt and 'horsepower'.

W.F.G.

J.H.C.

Contents

Foreword, v
Acknowledgements, viii
Chapter 1 Iron, 1
Chapter 2 Steel, 12
Chapter 3 Metal Finishing Processes, 23
Chapter 4 Machine Tools, 33
Chapter 5 Sources of Power 1, 49
Chapter 6 Sources of Power 2, 77
Chapter 7 Measurement, 88
Chapter 8 Craftsmen and Engineers, 98
Biographical sketches of famous engineers and ironmasters:
The Darbys of Coalbrookdale, 108
John Wilkinson, 110
Sir Henry Bessemer, 112
Sir William Siemens, 113
Sidney Gilchrist Thomas, 115
Joseph Bramah, 116
Henry Maudslay, 117
Sir Joseph Whitworth, 119
James Nasmyth, 121
Brown and Sharpe, 123
Thomas Newcomen, 125
James Watt, 126
Richard Trevithick, 129
The Hon C. A. Parsons, 131
George Stephenson, 132
Robert Stephenson, 134
I. K. Brunel, 136
Henry Ford, 138
Examination questions, 140–41
Name Index, 142
Subject Index, 145

Acknowledgements

A textbook of this sort draws a great deal upon the original scholarship of others, and the authors gratefully acknowledge their considerable debt to the writers and publishers of a number of authoritative works. To a select group of informative sources we are particularly indebted, and these are listed below.

W. H. G. Armytage. *A Social History of Engineering.* Faber 1961.

P. J. Booker, *A History of Engineering Drawing.* Chatto & Windus 1963.

D. J. Brandt, *The Manufacture of Iron & Steel.* E.U.P. 1960.

A. Burstall, *A History of Mechanical Engineering.* Faber 1963.

J. E. Carr and W. Taplin, *History of the British Steel Industry.* Blackwell 1962.

The Chartered Mechanical Engineer, especially articles written by L. T. C. Rolt (Oct. 1962 and May 1963), R. A. Hoare (Jan. 1963), J. Hoyland (Sept. 1965), F. R. McKears (Oct. 1965), A. E. Musson (April 1963).

S. D. Chapman, J. D. Chambers and T. R. Sharpe, *The Beginnings of Industrial Britain.* U.T.P. 1970.

G. D. H. Cole and Raymond Postgate, *The Common People*, University Paperbacks, 1968

H. W. Dickinson, *A Short History of the Steam Engine.* Cambridge 1939.

J. Gloag and D. Bridgewater, *A History of Cast Iron in Architecture.* Allen & Unwin 1948.

D. S. Landes, *The Unbound Prometheus.* Cambridge University Press 1969.

J. W. Roe, *English and American Toolbuilders.* Yale 1916.

L. T. C. Rolt, *Great Engineers.* Bell 1962.

H. R. Schubert, *History of the Iron and Steel Industry.* Routledge & Keegan Paul 1957.

C. Singer, *A History of Technology.* Oxford 1954-58.

We are grateful to the following for permission to reproduce copyright material: Joint Matriculation Board for sixteen questions from past examination papers.

Our thanks are due to Miss B. J. Ronchetti M.A., A.L.A. of the University of Birmingham Library, and to Miss H. Wells for help in preparing the original manuscript.

We are grateful to the following for permission to reproduce illustrations: Allen and Unwin Ltd, from Wolf, *A History of Science, Technology and Philosophy in the Sixteenth and Seventeenth Centuries, Vol. II*, fig. 54; B. T. Batsford Ltd, from L. T. C. Rolt, *Tools for the Job*, drawn by Max Millar, figs 40, 42; British Iron and Steel Federation, figs 5, 9, 16, 17, 27; Brown and Sharpe Manufacturing Company, figs 43, 45; Cambridge University Press, from H. W. Dickinson, *A Short History of the Steam Engine*, fig. 69; Constable and Company Ltd, from A. H. Gibson, *Hydraulics and its Applications*, fig. 56; John Cornwell, fig. 2; Crown ©, Science Museum, London, figs 18, 23, 34, 35, 37, 38, 48, 49a, 75, 76, 78, 89, 94; Mrs Curwen, fig. 51; E. A. Dixie, from *American Machinist 1913*, fig. 44; English Universities Press, from D. J. Brandt, *The Manufacture of Iron and Steel*, figs 1, 3, 13, 14, 15; Faber and Faber Ltd, from A. Burstall, *A History of Mechanical Engineering*, figs 29, 47, 92; John Freeman and Company, fig. 24; Harvard University Press, from Usher, *History of Mechanical Inventions*, figs 31, 32; Heinemann Educational Books/Schools Council, from *Industrial Archaeology of Watermills and Waterpower*, figs 52, 53, 55, 57; Institute of Mechanical Engineers, from E. G. Semler, *Engineering Heritage*, figs 11, 46; R. Kennedy, *The Book of the Motor Car, Vol. I*, fig. 50; A. F. Kersting, fig. 7; Manchester Photographic Society, fig. 21; Metropolitan Museum of Art, fig. 86; M.I.T. Press, from W. Parsons, *Engineers and Engineering*, fig. 59; Museum of Science and Industry of Birmingham, figs 39, 64, 68; Oxford University Press, from C. Singer, *A History of Technology*, figs 26, 28 and from Derry and Williams, *A Short History of Technology from Earliest Times to 1900*, fig. 87; Ronan Picture Library, figs 4, 12, 83, 85; Routledge and Kegan Paul Ltd, from Crowther, *Discoveries and Inventions of the Twentieth Century*, figs 73, 77; Sheffield City Museums, fig. 10; W. Tyzack, Sons and Turner Ltd, fig. 20; J. Wells, fig. 60; World Distributors Ltd, from Odle, *The Picture Story of British Inventions*, figs 62, 65, 80.

Chapter 1

Iron

MAN WITHOUT METAL—THE STONE AGE

For almost 99 per cent of the period of human habitation on the Earth man was without the knowledge of metals. We should not forget, however, that the men of the New Stone Age developed skills in the mining and shaping of flints which formed a foundation of technical knowledge from which later discoveries were to benefit.

THE EARLIEST METALS

In ancient history it is impossible to give precise dates, but we believe that by 3500 B.C. man was working in copper, silver, and gold. The ores of these metals were readily available in surface outcrops, and were easy to smelt and cast into moulds. Being very malleable they could be readily fashioned into simple jewellery, coins, and ornaments. The softness of these metals placed limitations on their use, and by about 1500 B.C. the mixing of copper and tin to produce bronze was being practised. This alloy gave a much stronger metal from which tools and weapons could be fashioned. Under the great civilisations which flourished at this time, as in Egypt and Sumeria, much progress was made in the techniques of metalworking.

The Discovery of Iron

There is evidence that iron was known as early as 2500 B.C., but its use did not become widespread until about 1400 B.C. The possession of this metal, and the ability to work it, played a great part in the military and engineering triumphs of the Greeks and Romans. Somewhat later the inhabitants of India, China and Northern Europe were able to demonstrate the metal's attraction for artist craftsmen.

Iron is one of the most important of the elements to be found in the Earth's crust, and minerals containing more than 20 per cent of iron are termed 'iron ores'. In the richest of the iron ores the iron is present in the form of an oxide of iron, and the original discovery of the metal probably took place when some of these ores were accidentally placed in very hot fires or even pottery kilns. When the wood used for fuel had been turned to charcoal (which is mainly carbon) a chemical reaction would then take place, resulting in the reduction of the iron oxides, i.e. removal of the oxygen, and the metal would then have collected at the bottom of the fire and solidified on cooling. It was soon realised that further heating and hammering of the iron after its removal from the furnace improved its quality.

The Iron Age in Britain

THE BOWL FURNACE

Iron was first smelted and worked in Britain about 450 B.C., the techniques having been introduced by Celtic immigrants from the continent. The type of furnace used was the bowl furnace which continued to be important until the early Middle Ages. It consisted of a shallow pit lined with clay which included provision for an air blast via a clay tuyere or flue.

THE PROCESS

Before smelting commenced the clay lining was hardened by burning charcoal in the bowl, after which the furnace was charged with alternate layers of charcoal and ore. The dome-shaped mound was finally covered with fine charcoal and pounded firm. Fire was introduced through the tuyere, and a gentle blast was started to remove moisture from the iron ore. A stronger blast was then applied and the oxygen in the air combined with the carbon from the charcoal to form carbon monoxide. This gas, passing through the iron ore, combined with the oxygen in it, reducing some of the ore to its metallic state. The reduction would take place at about 1200°C, well below the melting point of the iron (1528°C), and though the iron was never liquid, the crystals of iron united in a spongey mass or 'bloom' mixed with slag. The formation of the slag was caused by the combination of the silica present in the earthy portions of the ore (the gangue) and some of the ferrous oxides in the iron. Owing to the low furnace temperatures, the process was very wasteful, as much as 50 per cent of the iron being lost in the slag. The bloom produced by this type of furnace weighed about 3 kg, and before its use for manufacture it required prolonged heating and hammering. The resulting metal was a soft and ductile wrought iron with a carbon content of about 0.1 per cent. During the Roman occupation of Britain iron production expanded, although production methods remained basically the same apart from three innovations. The first of these was the practice of roasting the ore to remove moisture and increase its porosity before smelting took place. The second change involved a modification in the shape of the furnace, with the bowl furnace becoming straightsided. Finally, it became the practice to weld the small blooms together to form a much larger mass of iron.

Manufacture of iron continued in much the same way well into the sixteenth century, although the furnaces of the later period were producing blooms which often weighed up to 150 kg, and these would require the use of a water-powered hammer when they were beaten into wrought iron. On occasions after a high wind, the temperature of the furnace would be raised so that at one stage in the operation a fully molten iron would be in existence. Unfortunately this was too brittle to forge and it was only with the gradual emergence of a refining process that wrought iron could be produced from it. In time these early ironmasters must have learnt how to obtain molten metal without the freakish intervention of a strong wind and the way was then open for the development of a permanent furnace which could be tapped at regular intervals.

THE EARLY BLAST FURNACE

By 1400 German ironmasters were using 'Stückofen', a more effective kind of furnace some 5 metres high and 1 metre in diameter, to produce molten iron. Their natural successors, the first blast furnaces, were in operation in

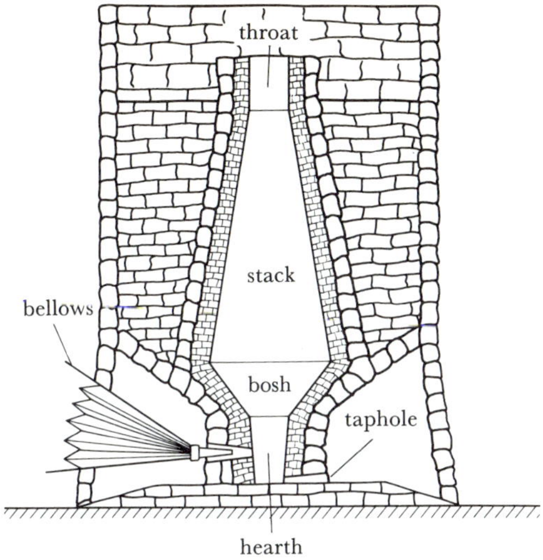

Fig. 1 Seventeenth century blast furnace

Britain by the late fifteenth century, and by the seventeenth century they had developed into the type illustrated (fig. 1). The height of the furnace was restricted to about 10 metres because the charcoal was unable to support a greater weight of ore without blocking the bottom of the furnace.

The iron ore and charcoal, together with limestone, were fed into the open top of the furnace, and production was continuous during the winter months of the year when sufficient water was available to turn the water-wheel which worked the bellows. This period of operation was known as a campaign, and the remaining part of the year was used for collecting charcoal and repairing the lining of the furnace.

It took some six tonnes of charcoal to produce a single tonne of iron. In England the industry was mainly concentrated in the Forest of Dean and the Weald of Sussex. Although coppices were actually planted for the iron trade the inroads made into the available timber led to restrictions being imposed on the amount felled for the making of charcoal in order to preserve supplies for shipbuilding. The increased size of the furnace, together with a more efficient blast which was usually supplied by water-powered leather bellows, raised the temperature inside the furnace sufficiently to produce a fully molten iron that collected at the bottom of the furnace. The iron was run off periodically and allowed to cool in a sand impression formed on the floor of the foundry. Where the resultant casting weighed over 500 kg it was known as a sow and smaller castings were termed pigs.

The iron from the blast furnace was much different from the product of the bloom furnace, and with a carbon content of between 2 and 5 per cent it was too brittle to be forged.

THE FORGE

While the output of some blast furnaces was used for casting directly into moulds, the majority of the iron produced underwent further treatment in the forge.

The first step was to remove the excess carbon in the iron by oxidation. This was undertaken in the *finery* where the pig iron was remelted and exposed to the decarburising effect of an air blast. The pasty mass which formed in the bottom of the hearth was collected on the end of a metal rod called a *ringer*.

To assist the decarburisation, the metal was subjected to continuous movement in the pool of slag which formed around it. Eventually the bloom was removed and forged into a bar about three feet long by using a water-powered hammer.

This bar or *ancony* was then transferred to another hearth or *chafery* where it was subjected to further reheating and hammering to improve the quality of the metal. The resulting product of the forge was wrought or bar iron. Unfortunately, the quality of this iron was largely inferior to that of certain continental countries which had access to deposits of high grade ore. In fact, before 1784, the British Admiralty, one of the largest users of iron, insisted that their anchors and anchor chains were to be made of imported iron from Sweden or Spain.

Coke as a Fuel

ABRAHAM DARBY I (1678-1717) AND ABRAHAM DARBY II (1711-63)

Although the production of iron was restricted for the reasons we have given, the demand for the metal steadily increased throughout the sixteenth and seventeenth centuries. The development of scientific ideas and the expansion of trade and industry (not forgetting the needs of warfare) were making demands which the iron industry, dependent on charcoal-burning blast furnaces, was unable to meet.

In the middle of the seventeenth century Derbyshire brewers were using charred coal (coke) as an improved method of drying their malt, and it is known that Abraham Darby had contact with the industry during his apprenticeship to a malt mill manufacturer. In 1709 at his ironworks in Coalbrookdale, Shropshire, he was using coke to replace charcoal as fuel for the blast furnace.

COKE

The coke was obtained by burning coal covered with clay and cinders to exclude air; gas fumes and tar were released, and the resultant product was coke, porous and non-volatile, the ideal fuel for iron making. By happy accident Darby used a 'clean' local Shropshire clod coal and avoided transmitting impurities through the direct contact of coke and iron. Other ironmasters not appreciating the need for a special coking coal or for a consistently high temperature must have experienced great difficulties where they tried the new method. Whatever the reasons little progress was made with coke-burning furnaces until Darby's son Abraham Darby II, began a serious study of the process in 1748. By selecting haematite ores, which have a low phosphorus content, he succeeded in improving the quality of the iron, and the use of coke for the smelting of iron became established.

THE BLAST FURNACE AFTER DARBY

The introduction of coke removed the restrictions that charcoal had imposed on the size of the blast furnace, coke being able to support a much increased load of ore, and as the size increased essential improvements in the technique of smelting iron were incorporated. In 1760 John Smeaton provided the necessary increase in the air blast using a blowing engine operated by a water-wheel, and later in the century one of Watt's first steam engines was used for this purpose by the famous ironmaster, John Wilkinson. By this stage coke smelting became as efficient a process as charcoal smelting. Darby's techniques could now be utilised to give the foundry an advantage over the forge.

In 1828 a Scot, J. B. Neilson, patented a method of preheating the air blast, although it is possible that Abraham Darby's grandson, Abraham Darby III, had been practising the same technique some fifty years before. Before Neilson, however, most ironmasters had taken considerable care to ensure that the air blast was cold. Preheating raised the efficiency of the furnace whilst the rate of production was significantly improved and fuel costs lowered. In 1857, E. A. Cowper patented a stove which stored and re-used the waste heat from the blast furnace gases. Many attempts were made to capture the gas which escaped from the top of the older blast furnace and re-use it. The first success in this venture occurred in 1849 when a Monmouthshire man, George Parry, devised the bell and cone device which, in a modified form, can still be seen on top of the modern blast furnace. The actual coking of coal since about 1763 had taken place in special circular ovens commonly referred to as beehive coke ovens. With the drive towards more efficient and economic methods of production these were replaced with the vertical coke oven. The new variety, introduced in 1881, conserved both gas and tar and could be adapted for automatic loading and clearing.

The Modern Blast Furnace

Blast furnaces have steadily increased in height until today many of them reach upwards of 30 m. They are steel plate structures lined with firebrick, or in some recent cases amorphous carbon blocks, and provision is made for water-cooling the lining to increase its working life. To meet the demands of

Fig. 2 Blast furnace at Llanwern steel works, Gwent

the modern furnace a large complex of auxiliary plant is required: coke ovens, ore-sintering equipment to make the soft native ores more suitable for smelting, and power stations to supply the electricity which operates the handling machinery. The furnace is charged with iron ore, coke, and limestone which promotes the formation of slag. These items are weighed and transported to the top of the blast furnace by automatic machinery. A bell-hopper arrangement prevents escape of gas from the furnace while loading takes place.

THE COWPER (OR HOT BLAST) STOVES

The gas produced inside the furnace is piped off from the top and led through a dust-removing plant into the base of the Cowper stove, of which there are usually three or more to a furnace. The inside is filled with a honeycomb of firebricks, the furnace gas is ignited at the base of the stove, and the heat produced is stored in the firebricks. When the bricks are fully heated the gas supply is diverted to another stove, and the air blast to the furnace is now passed over the heated bricks which raise the temperature of the blast to about 800°C. The process of storing heat in firebricks to be used later is called *regeneration*, and we shall meet it again in the chapter on steel.

THE PROCESS

The heated air blast enters the blast furnace via the bustle pipe and tuyeres, and as the reduction of the iron oxide in the ore takes place the molten iron sinks to the bottom of the furnace (the hearth) at a temperature of about 1600°C. The slag formed by the fluxing of the gangue with the limestone floats on top of the molten iron, and is drawn off through the slag notch at intervals determined by the quality of the ore being smelted. The tapping hole for releasing the molten iron is lower than the slag notch and is situated at 90° to it. The tap hole is sealed by a clay plug which is inserted mechanically, and to tap the furnace the plug is burnt away using an oxygen lance, allowing the iron to flow along a sand channel into large mobile ladles. Alternatively, the iron may be allowed to cool in impressions formed in a sand bed which are referred to as pigs, hence the term pig iron for the product of the blast furnace.

PIG IRON

The product of the blast furnace is an iron with a varying content of silicon, carbon, and other impurities; without further processing it is useless for manufacturing purposes. We must now examine the development of three important processes which converted pig iron into irons essential to an industrial economy.

Wrought Iron

As the demand for iron increased with the impetus of the Industrial Revolution, experiments were made to produce wrought iron of better quality than had hitherto been available when using low grade native ores. Once again Coalbrookdale figured significantly, for we know that three men from this area patented experimental furnaces for the production of wrought iron. The patent of 1766 was in the name of two brothers called Cranage, and that of 1783 in the name of Peter Onions. Neither patent came into commercial operation. The most successful work in this field was that of Henry Cort of Fareham in Hampshire, who had perfected his puddling process for converting pig iron into wrought iron by 1784. Cort's process took account of the need to use coal as a fuel instead of charcoal which was becoming increasingly difficult to obtain. The special importance of his furnace lay in the separation of the fuel from the metal which allowed him to use coal without contaminating the iron with sulphur.

The Puddling Furnace

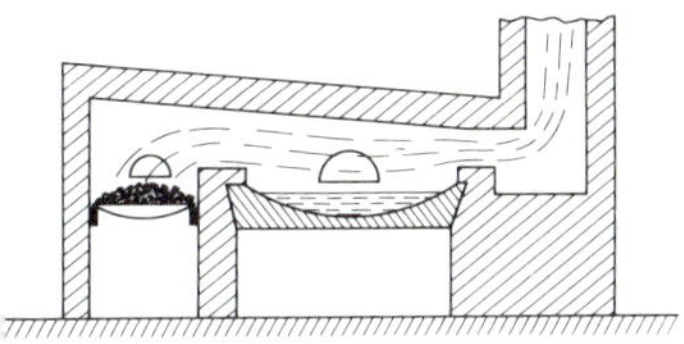

Fig. 3 Design of a puddling furnace

Figure 3 illustrates the principle of a puddling or reverberatory furnace. The heat from the fire is deflected (or reverberates) from the sloping roof on to the hearth, raising the temperature of the pig iron charge to about 1300°C. As the pig iron melts, iron oxide is added, and the impurities, silicon, phosphorus, manganese, and excess carbon are separated by oxidation, and form a slag. The slag and molten iron are stirred by the puddler with a long rod or rabble, and this promotes a chemical action to remove unwanted elements. When the process is complete, the puddler collects the pasty mass on the end of the rabble, and removes it from the furnace.

Fig. 4 Puddling furnace

The original Cort furnace had a sandy bottom to the hearth and the iron was heated to a pasty state. This was referred to as 'dry puddling'; an improvement was the 'wet puddling' process introduced by J. Hall, who used an iron oxide at the base of the hearth and increased the heat to melt the iron. Some slag was unavoidably removed with the iron, and much of it had to be removed by repeated folding and hammering. Finally, the wrought iron was passed through grooved rollers which slit the iron into bars, the remaining slag being present in the metal in the form of long fibres which gave wrought iron its characteristic fracture. Grooved rollers, a big improvement on the old slitting mill which required that the iron be first rolled into plates, were introduced by Henry Cort, who had been anticipated in their invention by a Swede, C. Polheim, in 1745. In 1789 Cort's involvement in a financial scandal lost him control of his patents and other ironmasters were given a heaven-sent opportunity to avail themselves of important new production techniques. Freed from what was described as 'the tyranny of wood and water' such industrialists built up new sites where the extraction of iron ore and coal and the production of iron were all contained within a single area.

Wrought iron is a soft and ductile metal, resistant to corrosion and readily welded by the hammer. Although largely superceded by mild steel for structural purposes, it is still used for the manufacture of crane hooks, chains, and railway draw bars. Under excessive strain it tends to stretch and distort rather than to snap suddenly, thus giving warning of impending failure.

Cast Iron

The ironmasters of the Middle Ages cast the product of their blast furnaces directly into moulds. With the size increase of the blast furnace after Darby II it became the practice to deliver the pig iron to the foundry which had become a separate industry. There the pig iron mixed with iron scrap was remelted in a small furnace or cupola, the invention of which was mainly due to the famous English ironmaster, John Wilkinson. Cast iron has a carbon content of 2.5 to 4 per cent, some of which is present in the form of graphite.

THE CUPOLA (OR FOUNDRY FURNACE)

The cupola is a small, simple type of blast furnace situated on the site of the foundry. It consists of a cylindrical steel plate container with an open top and a firebrick lining, and the blast is supplied by a small motorised fan. The furnace is charged with pig iron, steel scrap, and coke, as shown in the diagram (fig. 5), and a little limestone may be added to assist in fluxing. Little refining takes place in the cupola, but the quality of the product is controlled by careful grading of the scrap. The iron is tapped from the cupola into ladles and then poured directly into the prepared moulds. Cast iron is grey in colour, brittle, weak in tension, but relatively strong in compression. In recent years successful attempts have been made to improve the properties of this metal.

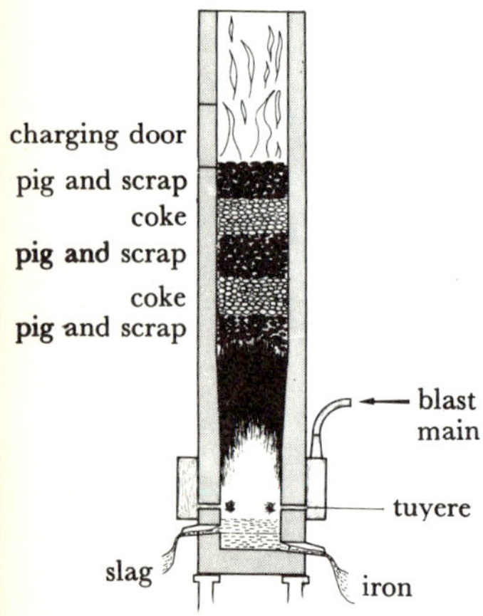

Fig. 5 Principle of the Cupola or Foundry Furnace

Fig. 6 Eighteenth century Cupola

Malleable Cast Iron

When iron is used for casting into small objects the rapid cooling which takes place renders the iron hard, extremely brittle, and very difficult to machine. The first recorded attempt to make cast iron malleable is credited to Prince Rupert, the Cavalier, who was granted a patent in 1671. The method involved heating either the cast iron or the pig iron in contact with iron ore for a long period in order to burn away some of the excess carbon. In 1722 a French scientist, Réaumur, used a similar process to improve the quality of pig iron, with some success. It seems that the process was little used during the eighteenth century, but after its rediscovery by Samuel Lucas in 1804 it was gradually adopted throughout the country. The production of malleable cast iron is now an important industry. Small cast iron parts which are subsequently malleableised are much cheaper than similar products in mild steel, and are only slightly inferior in strength.

Iron and the Industrial Revolution

The ability to produce cheap iron on a large scale was bound to have important repercussions in the economic life of industrial countries. Firstly, because of the technical developments which have been described, the iron industry in Britain became centralised in the coalworking areas. Regions like the Black Country, South Yorkshire, Clydeside, and South Wales, became centres of vast industrial concentration. Secondly, the great ironmasters, men like the Crawshays of South Wales, often became a byword for industrial enterprise, and they built up an empire which also embraced the mining of coal and iron ore and the sale of the finished product.

Perhaps most of all it was realised that there was now available a metal of amazing versatility, and the use of iron became associated with the most innovating engineers, architects, and businessmen. 'Iron Mad' John Wilkinson, more than any one of his age, exemplified the enthusiasm for iron. His more picturesque flights of fancy included the building of an iron chapel and an iron coffin for himself. But his iron water pipes for the city of Paris and his iron barge launched in 1787 were more positive demonstrations of his faith in the metal.

In an age where transport was assuming great significance, iron played a vital role. The Hon. John Byng, visiting Coalbrookdale in 1784, comments on the prosperity of the ironmaking there, and gives us the first description of an iron railway—'every cart belonging to this trade is made of iron, and even the ruts of the road are shod with iron'.

The first iron bridge, constructed at Coalbrookdale in 1778, was soon followed by many others. Telford, one of the leading civil engineers of his day, had already built five iron bridges before he made his famous proposal in 1801 to replace London Bridge by a single-arched bridge which would leap the Thames in a span of 180 m. The idea was finally abandoned, not through any structural difficulties, but because the approaches to the bridge were not fitted to cope with a heavy flow of traffic.

A cotton mill built in Manchester in 1801 for the firm of Phillips and Lee showed the great potentiality of iron as a medium for building. The mill,

Fig. 7 Iron bridge at Coalbrookdale

42 m long and seven storeys high, was supported by two rows of cast iron stanchions or uprights, and the floor itself was carried on inverted T cast iron beams some 4 m in length.

At the end of the eighteenth century architects began to use decorative ironwork to great purpose. At a time when there was no properly organised police force, iron railings were needed to protect the town house; iron balconies also became the vogue, and for both these purposes cast iron would be used. One of the greatest of British architects, John Nash, who practised at this time, has been described as 'quite merciless in his use of cast iron'. Under his direction iron pillars were constructed in the Greek Doric style and mistaken for stone, and at the Prince Regent's palace in Brighton he installed a staircase of cast iron in the Chinese manner, which, to the undiscriminating viewer, appears to be made of bamboo.

Unfortunately the later Victorians who followed Nash had all his enthusiasm, but they often debased the founder's skill with fussy and over-complicated designs taken from every period of artistic history. They used iron even for such objects as hall stands and garden furniture where it would obviously have been better to use a more suitable material such as wood or wickerwork.

During the years 1801 to 1851 the population of Britain doubled, and since cheap steel was not yet available the demand for iron grew apace. The patenting by Neilson of preheated blast, which has been described in this chapter, allowed considerable economies of cost to be made, and the fuel needed to produce one tonne of iron was reduced from eight tonnes to five. It also allowed deposits of iron ore and coal which had previously been considered unusable to be exploited. Output was greatly improved as is shown by the production figures for Scotland: 37,000 tonnes in 1830; 200,000 tonnes in 1840.

Iron produced at low cost made possible an age of machinery which was fundamental to economic development. Already during the eighteenth century it had been the practice to replace wooden machine parts with iron, and this process was carried further after 1800. Maudslay, for example, built his lathes entirely of iron, and under his successors, Nasmyth and Whitworth, machine tools of iron became even more robust and solid.

The most ambitious applications of iron, however, were seen on the railways. One of the great railway schemes which captivated the imaginations of the majority of people at this time was the laying of the Manchester to Liverpool line, where George Stephenson was the chief engineer. It was found necessary to build altogether sixty-three bridges, and on many of these a normal arch could not be used because the line had to be level. Stephenson's first biographer, Samuel Smiles, described how, in these circumstances, Stephenson 'employed simple cast iron beams . . . by which he safely bridged gaps of moderate width, economising headway and introducing the use of a new material of the greatest possible value to the railway engineer'.

The most impressive constructions of the railway age were the great London terminals, where cast and wrought iron were used to give status and spaciousness to the power of steam. At Euston the Doric arch which formed the main entrance was complete with iron gates and railings, whilst at St Pancras, built in 1868, a huge glass roof vaulted the lines, almost 75 m in span and entirely supported by delicate iron trusswork.

Additional Reading

T. S. ASHTON, *Iron and Steel in the Industrial Revolution*, Manchester University Press 1951.

D. J. BRANDT, *The Manufacture of Iron and Steel*, E.U.P. 1960.

W. K. V. GALE, *Iron and Steel*, Longman 1969.

H. R. SCHUBERT, *History of the British Iron and Steel Industry*, Routledge & Kegan Paul 1957.

Chapter 2

Steel

Steel is an alloy of iron and carbon in which all the carbon is chemically combined with iron. The percentage of carbon which can be absorbed is limited, varying between 0.01 and 1.5 per cent. Within this small range the properties of steel change from those of mild steel at the lower end to those of carbon tool steel at the upper end. In this chapter we shall examine the methods of steel manufacture in the order of their discovery or invention, but it should be clearly understood that the use of steel for structural purposes waited upon the invention of special techniques which did not appear until the middle of the nineteenth century when the Industrial Revolution was well into its stride.

The Early History of Steel in Britain

Although steel was produced in Asia Minor as early as 1400 B.C. its introduction into this country is placed not earlier than the third century B.C. Two methods of production were used, the most important one being the surface carburisation of wrought iron.

Carburisation was accomplished by prolonged heating of the iron in charcoal, carbon from the charcoal being absorbed into the surface of the red-hot metal (a similar process to this is still used today for case-hardening small mild steel tools in the school workshop). The alternative method was to smelt the steel directly from manganese-bearing ores containing very few impurities. This method of producing steel remained in use up to the eighteenth century, and the product was referred to as natural steel. Output was, of course, restricted to localities where suitable ore was available.

Hardening steel by quenching from red heat appears to have been practised from the earliest times, but there is no evidence of the hardened steel being tempered to reduce brittleness prior to the Roman occupation. As steel was difficult and expensive to produce the technique of 'steeling' was developed by the Romans. This involved welding a steel cutting edge onto a wrought iron body.

The manufacture of steel in this way continued with little change up to the beginning of the seventeenth century. By then the demand for steel had grown to such an extent that it could only be satisfied by large importations from the continent. Home produced steel, with the exception of the small amount of natural steel, was inferior to imported steel. It was extremely variable in quality owing to the low-grade phosphoric ore from which it was produced. In an effort to improve the quality of British steel, William Ellyot and Mathias Meysey were granted a patent in 1614 for the conversion of iron into steel in a coal-fired reverberatory furnace. This process became known as *cementation*.

THE CEMENTATION PROCESS AND BLISTER STEEL

Although the original object of the patent was to make steel from domestic wrought iron, it rapidly became the practice to import Swedish bar iron to improve the quality of the steel.

Fig. 8 Cementation furnace at Derwentcote, County Durham

The process was carried out in conical furnaces of the type shown in fig. 8. This particular furnace housed two cementation 'chests' each capable of converting about 5 tonnes of iron. The chests were made from a refractory material such as sandstone. The wrought iron was packed into the chests, between layers of crushed wood charcoal, to within 10 cm of the top. The chests were sealed by filling the remaining space with 'wheelswarf', a mixture of sandstone and steel particles collected from the grinding shops of the cutlery trade. When heated, the wheelswarf formed an airtight seal to the top of the chest.

When the packing of the chests was completed, the entrance to the furnace was sealed off with bricks and clay and then a coal fire was ignited beneath the chests. The 'heat' continued from 5 to 9 days depending on the quality of the steel that was being manufactured. On completion of the firing process, a cooling period of seven or eight days was necessary before the bars of steel could be removed from the chests. Thus each cycle of the cementation process occupied some 15 to 18 days from start to finish.

SHEAR STEEL

While blister steel was suitable for general smithy work and the production of cheap cultery, it was unsuitable for the manufacture of articles which required a keen and strong cutting edge such as razors and lancets. To produce a steel of more uniform quality, bars of blister steel were broken into small lengths and bound together to form a 'faggot'. The faggots were brought to a welding heat and forged into a single bar. This was known as shear steel. A further improvement in quality could be made by repeating the operation to give 'double shear' steel.

The production of steel by the cementation process reached its zenith in the 1860s when approximately 80,000 tonnes were being produced annually in the Sheffield area alone. Production started to decline towards the end of the nineteenth century as the use of blister steel in the crucible process was superseded (see below). The last cementation 'heat' in Britain took place at the factory of Daniel Doncaster Ltd. of Sheffield in 1951.

Crucible Cast Steel

BENJAMIN HUNTSMAN (1704-76)

The need for steel of a better grade than could be obtained by the cementation process inspired a Yorkshire clockmaker, Benjamin Huntsman, to investigate methods of producing steel more suitable for the manufacture of clock springs. He realised that only a thorough dispersion of the carbon throughout the steel would provide the quality he sought, and by 1740 he had perfected a method of refining blister steel by melting it in a closed vessel, a crucible, made from a refractory material. One of his most difficult tasks was the production of a crucible which could be handled at red heat when containing molten steel. It soon became apparent that 'Huntsman steel' was a superior product, not only for the manufacture of clock springs, but also for the cutlery trade. After the secret of his process leaked out other firms quickly adopted his method, and by the 1780s Sheffield had established its reputation for the production of high quality crucible cast steel.

While it would be wrong to credit Huntsman with the invention of crucible steel, he must certainly take the credit for its rediscovery and its introduction on a commercial scale into Western Europe. Where Huntsman thought in terms of individual crucibles, later steelmakers developed methods of heating and pouring crucibles *en masse*. The firm of Krupp in Germany was able to produce ingots weighing several tonnes and by 1869 the Vickers company in Britain coordinated the output of 672 crucibles at a time.

The Crucible Process

The Huntsman process changed little throughout its long life (almost 200 years) as the main source of high quality tool steel. Although Huntsman used blister steel originally, it gradually became the practice to use Swedish bar iron further to improve the quality. The charge of iron, broken into small pieces, together with crucible steel scrap and a small amount of charcoal, was carefully weighed before placing into the preheated crucible which stood inside the furnace surrounded by a coke fire (fig. 9). After the crucible had been covered, the air draught increased the furnace temperature to around 1550°C when the charge melted and was thoroughly mixed by the resulting agitation. This thorough mixing was the main objective of the crucible process; carbon content and quality being controlled by the careful selection and weighing of ingredients.

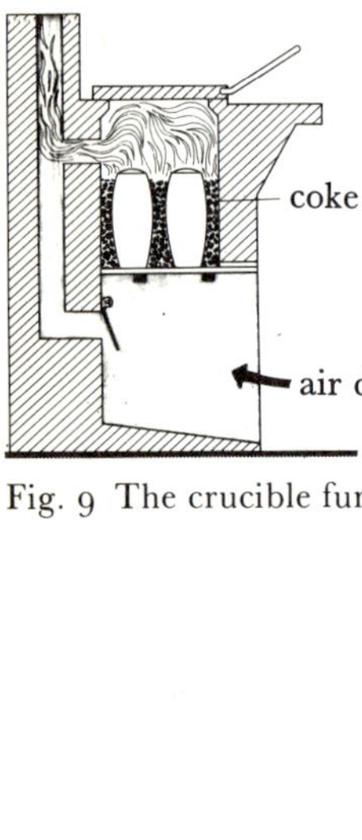

Fig. 9 The crucible furnace

When the 'melt' was completed, after some 3 to 4 hours, the crucible was lifted from the furnace, the slag removed, and the contents teemed into an ingot mould. The weight of the charge was around 27 kg.

Fig. 10 Crucible steel furnace floor (aerial view) Abbeydale, Sheffield

The photograph (fig. 10) of the crucible furnace, preserved at Abbeydale Industrial Hamlet, shows clearly the crucibles (pots) and the pot-holes on the left of the picture, while on the right can be seen the split ingot moulds with one mould standing in the casting pit ready for teeming.

Crucible cast steel had a carbon content of between 1.0 and 1.5 per cent, and was eminently suitable for the manufacture of high quality cutting tools of all kinds. It could be annealed (softened) by heating it to a cherry red and cooling slowly in lime, after which it could be filed. It was hardened by quenching in water from a cherry red and required tempering (further heating to reduce hardness) in order to retain the cutting edge.

The crucible process was gradually superseded by the electrical refining processes (see the end of this chapter) during the first half of this century. Although one or two furnaces were relit during the 1939–45 war, the process was virtually obsolete by that time.

Mild Steel

SIR HENRY BESSEMER (1813-98)

In 1856 a paper entitled 'On the Manufacture of Malleable Iron and Steel without Fuel' was read to the British Association for the Advancement of Science. The author was Henry Bessemer who had already many discoveries to his credit, the most important of which was the manufacture of 'gold' powder by a cheap mechanical method. He now turned his attention to the production of iron, and found the conventional puddling furnace troublesome and messy. Bessemer described to the British Association an object which he called the converter, which was to prove capable of producing cheap steel in large tonnages. Mild steel with a carbon content below 0.3 per cent was fairly soft and ductile, and was easily machined and forged. At the same time it possessed good wearing properties, and was admirably suited for making boiler plate, railway lines, rolled steel beams and so forth, which were so much in demand at that time of rapid industrial expansion.

Research along directly parallel lines had been proceeding under James Nasmyth in Britain and in the United States under a small ironmaster, William Kelly. Bessemer's public utterances anticipated any discoveries Nasmyth might have made in this field whilst the technical superiority of the Bessemer converter itself undermined the value of patents already taken out by Kelly in the U.S.A. Thus it was Bessemer who by acquiring British and American patents was able to give his name to the famous tilting converter. The confusion arose when Bessemer insisted on calling his metal steel. He had in fact produced a low carbon (mild) steel with properties much different from those of the high carbon product of Huntsman process. The other great drawback of Bessemer's method was that excess oxygen caused blow holes in the metal. It was Robert Mushet, a metallurgist and owner of a small works in the Forest of Dean, who showed that the oxygen could be absorbed by adding an alloy of manganese and iron. This alloy could also be used to make the final adjustment to the carbon content of the metal.

THE BESSEMER CONVERTER

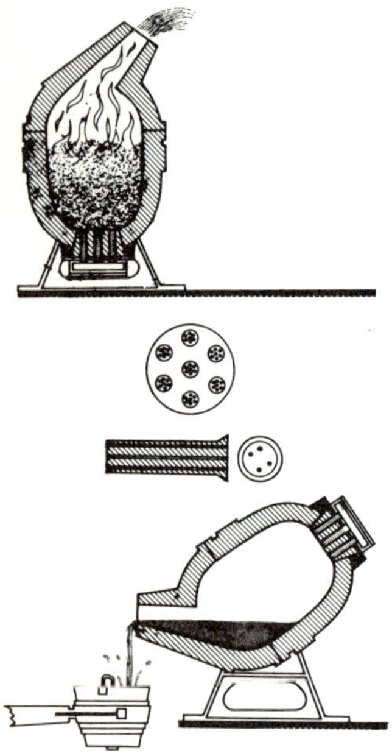

Fig. 11 Original drawings of the Bessemer converter

The converter is a large open-mouthed refractory-lined steel container mounted on trunnions which enable it to be rotated, and through which passes the air blast to the bottom of the container (fig. 11). The preheated converter is charged with molten iron from the blast furnace, the air blast at a pressure of about 25 to 30 lb per square inch (1.75 to 2.10 kg/cm^2) is turned on, and the converter is rotated to the vertical or 'blowing' position. Impurities in the metal are oxidised by the oxygen in the air blast, and the heat generated by these reactions is sufficient to keep the metal in a liquid state. The required amount of carbon is then added to the melt in the form of ferro-manganese, and the whole operation is completed in about twenty-five minutes. Finally the converter is rotated to empty its contents into ladles and then into ingot moulds, prior to its conversion into mild steel sheet, bar or strip.

In his original experiments with the converter Bessemer had made accidental use of iron smelted from haematite (acid) ores, and it was soon realised that it was unsuitable for converting iron ores containing a high percentage of phosphorus. By the time a solution to this problem was found confidence was lost, and the converter was slow to find favour in this country. In other countries Bessemer's process was quickly adopted and the world production of mild steel expanded rapidly. The largest producer was the United States, where ironmasters like Andrew Carnegie were quick to appreciate its potentialities. In fact, by 1900 there were thirteen places in the U.S.A. called Bessemer.

Fig. 12 The Bessemer process

The Open-hearth Furnace

SIR WILLIAM SIEMENS (1823-83)

The open-hearth furnace was developed by William Siemens, a member of a famous German family who were amongst the pioneers of electrical engineering. Siemens came to England as a young man and, after notable work in other branches of engineering, concentrated upon the improvement of the reverberatory, or puddling, furnace. Two German technicians Lohage and Bremme had already proved during the 1840s that the puddling furnace could be used specifically for steelmaking. Their innovation made something of an impact in France where it cut costs to £10 a ton. The major drawback was the difficulty in controlling carbon content during production and this prevented the system from being widely adopted.

The main feature of Siemens's work involved the regeneration, or re-use, of the heat which, in the reverberatory furnace, was allowed to escape up the chimney. He accomplished this by using checker chambers through which the waste furnace gases were directed. The firebricks from which the chambers were built absorbed heat from the gases, and when the bricks were fully heated the hot gases were directed into a second chamber. The cold draught to the furnace was now directed through the hot chamber so that the air was preheated before reaching the hearth. The two checker chambers were used alternately, one heating the air draught while the other absorbed heat from the waste gases. Preheating the air in this fashion raised the hearth temperature sufficiently to bring iron to its fully molten state, a task impossible in the reverberatory furnace with its maximum temperature of 1300°C.

The first patent for a regenerative furnace of the type described was taken out by William's brother, Frederick, in 1856. The following year Edward Cowper patented a similar regenerative system for the blast furnace (see chapter 1).

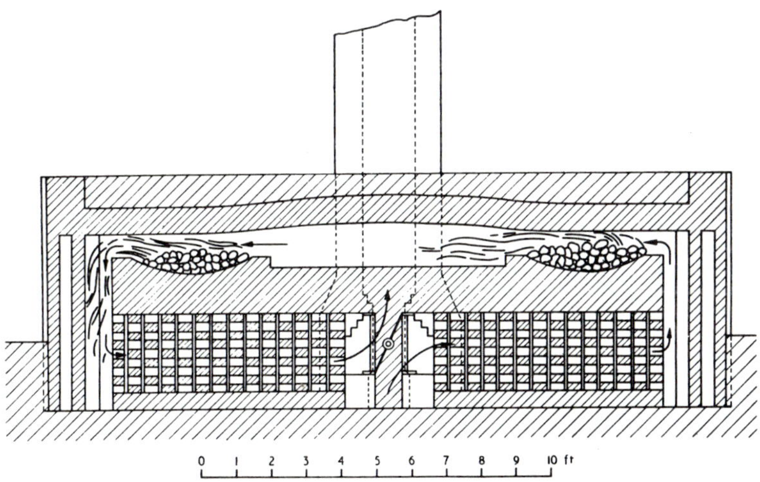

Fig. 13 Siemens's original furnace

Siemens's original furnace (fig. 13) incorporated two coal fires situated at opposite ends of the hearth, and it was first used for heating wrought iron billets, and later for melting crucible steel, reducing fuel costs by over 70 per cent. To overcome the problems caused by the products of combustion being carried over the hearth, Siemens invented the gas producer (fig. 14) which made a coal gas from cheap low-grade coal. The gas and the air were preheated by passing through separate checker chambers, meeting at the edge of the hearth where they ignited to produce a very hot flame.

In France the brothers Pierre and Emile Martin used a Siemens regenerative furnace successfully to make a steel of controlled carbon content by melting a charge of pig iron and wrought iron scrap, a process which they

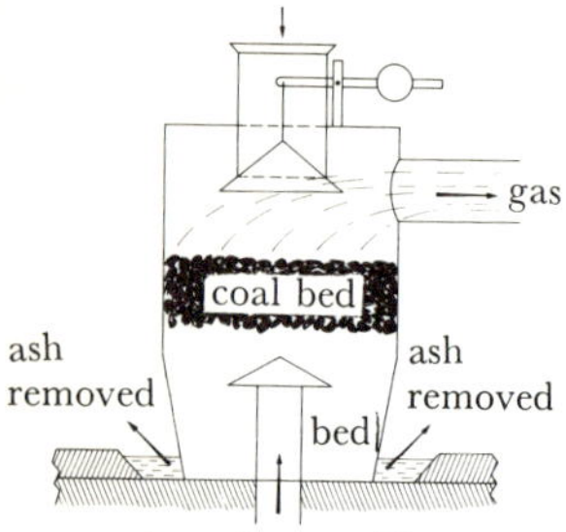

Fig. 14 Principle of the gas producer

patented in 1865. Siemens was also anxious to use his furnace to make steel from pig iron, but his first attempts were unsuccessful as the prolonged heating destroyed the furnace lining. The French method of charging the furnace with pig iron and scrap was eventually adopted and the process became established as the Siemens-Martin open-hearth furnace.

Siemens perfected the open-hearth furnace in 1867, at a time when the majority of British ironmasters were disgruntled by the failure of the Bessemer converter to refine phosphoric iron ores. They were eager to adopt this alternative method of producing mild steel in large quantities, even though low phosphorus iron ores were still required for the open-hearth furnace. One attraction of the open-hearth furnace was its ability to take up to 50 per cent of scrap wrought iron in the charge (in fact scrap was essential to the process). Britain was at this time the only country with sufficient quantities of scrap available to make the process economic. For these reasons the Siemens-Martin furnace became the major producer of mild steel in this country.

A costly defect of the new steel furnaces was their inability to refine pig iron produced by phosphoric iron ores. For a while British steel makers coped with the problem by importing high grade ores, but the real answer lay in the solution of a fundamental chemical problem.

Basic Mild Steel

SYDNEY GILCHRIST THOMAS (1850-85) AND PERCY GILCHRIST (1851-1935)

Sydney Thomas was a clerk in a London police court and an amateur chemist. In collaboration with his cousin, Percy Gilchrist, a chemist employed in a South Wales iron works, he investigated the problem of refining high phosphorus iron ores using the Bessemer process and by 1878 they had succeeded in their task. Thomas realised that the solution could be found by adding lime, but unfortunately lime attacked and ruined the silica lining of the converter. It was at this stage that he made his fundamental discovery of the dolomite lining. Dolomite, a mineral compound of calcium and magnesium carbonates resisted the action of the limestone which could now be added, enabling the phosphorus to be removed in the form of slag which floated on the surface of the melt. This became known as the 'basic' process to distinguish it from the 'acid' process of the original Bessemer converter in which the furnace was lined with silicate.

The Thomas converter, as it became known on the continent, was quickly adopted for use with high phosphorus iron ores, of which there were large deposits in eastern France and western Germany. For the reasons mentioned earlier, however, it failed to find much favour in this country.

The more attractive proposition to the British ironmasters was the logical extension of Gilchrist Thomas's work in its application to the open-hearth furnace. This was accomplished in 1879 through the work of two Frenchmen, Pourcel and Valrandin, and enabled 'basic' lined open-hearth furnaces to cope with the smelting of low-grade British ore.

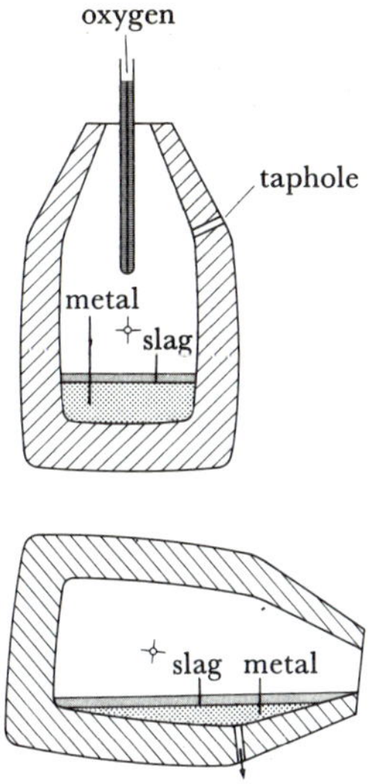

Fig. 15 The Linz-Donawitz process

BASIC OXYGEN STEELMAKING (BOS)

Until the early 1950s, over 90 per cent of the world's tonnage of mild steel was produced by the Bessemer and the open-hearth processes. Bessemer steel contains a small percentage of nitrogen introduced by the air blast, which for many uses of mild steel is quite acceptable. It does, however, make it unsuitable for use in mild steel sheet used for modern deep drawing methods of producing car bodies. This presented a serious problem to the countries faced with the postwar boom in car production and dependent on the Bessemer converter for their mild steel tonnage. Austria, where availability of low-phosphorus ores had made the Bessemer process predominant, was particularly affected and in the towns of Linz and Donawitz a modified Bessemer process (fig. 15) was perfected which produced nitrogen-free steel.

In the Basic Oxygen process the air blast holes through the base of the converter are replaced by a water-cooled lance which is inserted into the mouth of the converter to within a few feet of the molten charge. Pure oxygen is blown through the lance onto the surface of the charge at a rate of 860 m^3/min. This causes turbulence in the metal and an increase in temperature which burns out the unwanted impurities.

As well as producing nitrogen-free steel, this process has another great advantage over the old Bessemer converter in that the charge can consist of up to 30 per cent of steel scrap. In the largest modern converters over 350 tonnes of metal can be refined in one charge in about 40 minutes with high control over quality. As a result of these improvements, the Basic Oxygen process is rapidly replacing the open-hearth furnace as the major method of producing mild steel.

The word 'Basic' refers to the non-acid refractory lining of the converter which is better able to withstand the fierce temperatures and chemical reactions that occur during the process.

Electric Steelmaking Furnaces

As the knowledge of electricity increased throughout the nineteenth century, inventors experimented with ways of using the heating effect produced by an electric current. Siemens in 1879, and Cowles in 1886 each constructed experimental furnaces which were capable of melting iron by using electricity, and the Cowles furnace was used commercially for preparing alloys of aluminium and copper.

The important work on the electric furnace, however, was accomplished by Henri Moissan in 1892, working at the Sorbonne in Paris. He showed that the high temperatures which could be produced in the electric arc furnace (up to 4000°C) could readily accomplish the task of separating the elements of manganese, tungsten, molybdenum, chromium, vanadium, and cobalt, which hitherto had only been produced in small quantities with great difficulty. He showed that his furnace was also suitable for alloying these elements with steel.

THE ELECTRIC ARC FURNACE

This furnace, based on Moissan's discovery, was developed by Paul Héroult in France during the 1880s. The process is dependent on the fact that when

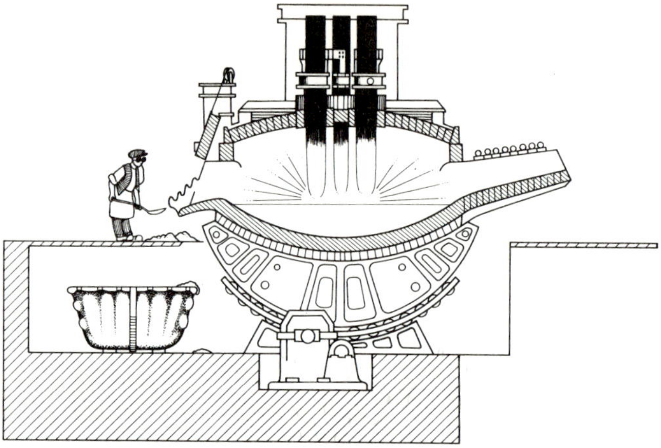

Fig. 16 An electric arc furnace

an electric circuit is broken a spark will leap across a small gap in the circuit. This spark can generate an extremely high temperature, over 3000°C, which is well above the melting point of steel. If the gap exists between two pieces of carbon (used because it will not melt) a steel object placed in the gap will be quickly melted.

In the modern electric arc steel furnace the heat is generated by an arc struck between three graphite electrodes and the steel to be melted. The electrodes are connected to an alternating current transformed to give about 30,000 amps at 100 volts. The furnace is charged with selected steel scrap and the graphite electrodes are lowered mechanically until a spark is struck between them and the charge. As the steel melts the graphite electrodes are lowered to maintain the spark.

Additions of lime, fluorspar and iron oxide are made which combine with the impurities in the metal to form a liquid slag. Samples are taken and analysed to check the composition of the steel and once the correct composition is achieved the slag is raked off the surface of the charge into waiting ladles. The furnace is then tilted and the charge poured into a teeming ladle suspended from an overhead crane.

The process was originally used for producing special high-quality alloy steels but today it is also widely used for making large tonnages of the more common steels. A large electric arc furnace can produce 150 tonnes of steel in 4 hours.

THE HIGH FREQUENCY FURNACE

In the 1890s Ferranti experimented with the use of another electrical phenomenon, induction, to melt the steel. The Swedish engineer, Kjelin, developed a successful induction furnace which forms the basis of the modern process.

If an alternating current is passed through a coil of wire, heat will be generated in any metal object which is placed inside the coil. This is the principle of the induction furnace, which has replaced the old crucible furnace. The furnace (fig. 17) consists of a refractory lining around which is passed a copper tube whose ends are connected to an alternating current generator supplying electricity at about 1000 cycles per second, and giving 400 amps at 2000-3000 volts. Water is passed through the copper tube to keep it cool.

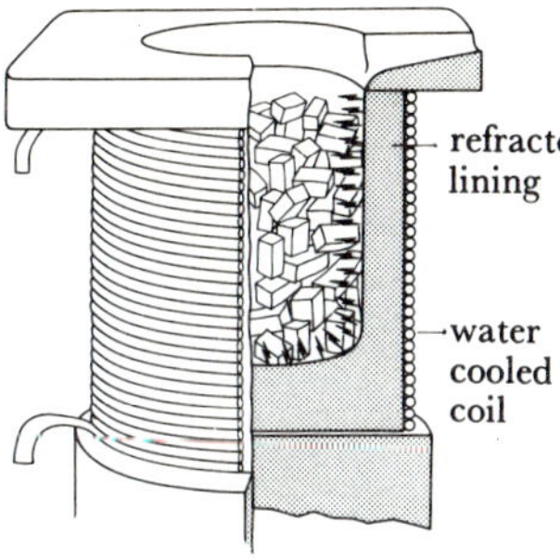

Fig. 17 Electric induction furnace

When the charge of carefully selected steel scrap is placed in the furnace and the current switched on the heat generated in the charge is sufficient to melt it. As the heat is generated within the metal itself there is no risk of contamination as there was in the crucible furnace due to seepage of gas through the walls of the crucible. As in the crucible furnace only melting takes place but the induced currents set up a turbulence in the charge which ensures a thorough mixing.

The capacity of an induction furnace can vary from a few kilos to ten tonnes, and the furnace can also be used for the production of high quality alloy steel for use in such demanding work as gas turbine blades. Like the arc furnace it is expensive to install and operate.

Alloy Steels

The alloying of steel to improve its properties was well advanced by the end of the nineteenth century. One of the early discoveries was that of manganese steel in 1882 by Sir Robert Hatfield of Sheffield. This steel, containing about 1 per cent carbon and 13 per cent manganese, becomes exceedingly tough when heated to 1000°C and quenched. It also work hardens rapidly and is difficult to machine with normal tools.

The discovery of nickel in 1751 had few immediate consequences because of the metal's brittleness. But it was eventually found that nickel had considerable value when included in alloys, and in 1889 a Glasgow man, James Riley, published his researches on nickel as an alloy in steel. The findings were taken up by the United States Navy, who first used nickel steel for armour plating, but in the course of time its use spread beyond the armaments industry into automobile and aircraft manufacture. When nickel is accompanied by controlled quantities of chromium and carbon, it is possible to produce a steel both stainless, tough, and ductile.

Another notable achievement was the manufacture of high-speed steel for cutting tools, produced in Britain by Robert Mushet under conditions of such secrecy that the working formula was never divulged. Wide recognition was achieved by the cutting steel developed in the United States by F. W. Taylor, Maunsel White and others, containing 0·7 per cent carbon, 18 per cent tungsten, and between 4 and 6 per cent chromium. It was capable of a cutting speed five times greater than that of carbon steel tools, and retained its cutting edge at red heat.

The properties imparted to steel alloyed with certain elements are given in the following table:

Alloying elements	*Effect*
Manganese	Gives exceptional hardness and ductility, makes steel non-magnetic. Imparts work hardening properties.
Tungsten	Increases hardness and strength at high temperatures.
Molybdenum	Increases elasticity and strength at high temperatures.
Chromium	Increases hardness, resists wear and corrosion.
Vanadium	Toughens and strengthens. Improves resistance to fatigue.
Cobalt	Increases hardness and retention of magnetism.
Nickel	Increases tensile strength, toughness, hardness and resistance to fatigue.

The metallurgical practice of alloying depended for its long-term success on the formulation of an exact scientific approach. Two milestones along this road were the discovery of the micro-structure of steel in 1864 and also that of X-ray diffraction – first applied by Bragg to crystalline solids in 1912.

The Steel Boom

Once steel had been made available in large tonnages it was soon accepted as the most useful and versatile engineering metal. In terms of cost, the combined effect of the work of Bessemer, Siemens and Gilchrist Thomas was to reduce the price of steel by 80–90 per cent. In 1870 world output was half a million tonnes, but in 1914 the output was 28 million tonnes, and by 1971 the U.S.A. was producing 107 million tonnes and the U.S.S.R. 118 million tonnes.

Although puddled iron remained the dominant metal in Britain as late as 1885, the technological change which transformed life at the end of the nineteenth century owed much to steel. The first rail of Bessemer steel was laid in 1865 and during the next 30 years, steel production benefited greatly from the world expansion in railway building. Girders and concrete reinforced by steel rods also opened up new possibilities in building and civil engineering. In mining and shipbuilding, steel wire ropes proved of great value and giant industries of the twentieth century manufacturing cars and domestic appliances depend very heavily on supplies of steel.

Increased output of this metal strengthened the sinews of war and contributed greatly to the arms race which has proceeded almost unabated since 1870. Steel helped to change the comparative economic position of the nations. For example, the arrival of Germany and the United States as world powers was signalled when they outstripped Britain as steel producers in the years before the first world war.

In the 1930s it was the turn of Russia, and the Communist government pushed steel production to the forefront of its five-year plans. During the Russian Revolution many of the leaders changed their names; one of the shrewdest took the name of Stalin, which in Russian means 'man of steel', a significant commentary on the reputation of this metal in modern times.

Additional Reading

D. J. BRANDT, *The Manufacture of Iron and Steel,* E.U.P. 1960.

J. E. CARR and W. TAPLIN, *History of the British Steel Industry,* Blackwell 1962.

N COSSONS, *The B.P. Book of Industrial Archaeology,* David and Charles 1975

SHEFFIELD CITY MUSEUM, Information Sheets Nos 1–9.

Chapter 3
Metal Finishing Processes

In the previous chapters we examined the production of ferrous metals. Now we must study the different methods of turning these metals into a marketable commodity.

The techniques of casting, forging, and wire drawing are probably as old as the history of metal itself, and we can still find many examples of the skill shown by early metalworkers in the use of these processes. In this chapter we shall trace the development of these methods, particularly under the impetus of the Industrial Revolution, and we shall also look at the techniques which were the product of the increasing demand for iron and steel during the last two hundred years.

Casting

IN BRONZE

The history of casting began in the Bronze Age. By 2000 B.C. bronze was being cast into rough clay moulds before it was finally shaped and hardened with the hand hammer. Axes, rasps, and cold chisels were amongst the articles produced in this manner. The bell-founder of Anglo-Saxon Britain was skilled in the production of large castings, using one-off moulds which were prepared on the site of the building where the bell was to hang. The skill acquired by the bell-founder was later used to produce cannon which came into general use in the fourteenth century.

IN IRON

The casting of iron only became possible with the introduction of the blast furnace in the late Middle Ages, and though cannon were being cast in iron in this country by the late fifteenth century, bronze was still preferred as a more reliable metal.

CASTING OF CANNON

By the middle of the sixteenth century the casting of iron cannon and shot was well established in this country. In 1543 Henry VIII ordered the construction of a foundry at Buxted in Sussex for this purpose. The director of the project was a parson, William Levett, who was fortunate in obtaining the services of Ralph Hogge, a skilled ironmaster, to supervise the work. The cannon produced by Hogge very quickly established a reputation for quality and reliability, and until displaced by Swedish cannon in the seventeenth century they were much sought after by the continental powers.

The casting of cannon involved the use of a three-part loam mould, the loam being reinforced with iron bars. The first part had an internal reproduction of the exterior of the cannon, including the trunnions (pivots), the second part was a mould of the breech, and the third was a core which fitted into the first part to provide the space required by the bore. The molten iron was run from the blast furnace directly into the moulds.

Fig. 18 Seventeenth century cast iron fireback by Richard Lenard

OTHER EARLY USES

Cast iron was also used in the sixteenth century for the production of fire-backs, fire-dogs, grave-slabs, and domestic ware such as pots, kettles, and pans. In industry cast iron anvils and hammers were used, and in the seventeenth century cast iron rollers for crushing sugar cane and for metal rolling were also being produced.

LATER DEVELOPMENTS

Before the beginning of the eighteenth century, moulds were made of clay or loam, and the metal was poured directly from the charcoal blast furnace. Abraham Darby I obtained a patent in 1707 for casting iron pots in sand, and was probably the originator of the modern method of casting in sand, using two or more moulding boxes.

By the latter half of the eighteenth century, Britain had established a commanding lead in the production of iron castings. The Carron foundry in Stirlingshire was particularly famous for its cannon which became known as 'carronades'. They were short, light and reliable, in an age when gun bursts were frequent, and fired a 30 kg shot. In 1777 Abraham Darby III erected the first cast iron bridge at Coalbrookdale, the castings being made in open sand moulds with the metal poured directly from the blast furnace. The parts of the bridge fitted together without use of bolts or rivets, and the fact that the bridge is still standing is sufficient tribute to the soundness of its construction.

By the mid-nineteenth century the first mould-making machine had appeared. It consisted of a table on trunnions which allowed the bottom box (or drag) and pattern to be rotated through 180 degrees prior to ramming the top box (or cope).

James Nasmyth made a notable contribution to safety in the industry when he introduced the worm-geared tilting pouring ladle which eliminated the commonest source of accidents.

Forging

Forging in its original concept applied to the work performed in the bloomery (see chapter 1). The finished bloom of wrought iron was the raw material of the blacksmith who, by further heating and hammering, could work up the iron into the shape he required.

In Celtic Britain the smith used a small stone anvil and probably a stone hammer, and that he developed considerable skill despite his primitive equipment is shown by the examples of ironwork belonging to this period. The techniques of the blacksmith steadily improved so that by the time of the Roman occupation chain-making, the working of iron into sheets, and the hardening and tempering of steel were included among the blacksmith's skills. Anvils and sledgehammers of iron were produced and the majority of the hand tools which we associate with the blacksmith had been developed.

In the period of stagnation which followed the Roman occupation the work of the smith was still in demand, and during the Anglo-Saxon period his skill in the forging of weapons of high quality increased. The smith was still predominantly an armourer until well after the Norman Conquest. Gradually, however, domestic ironware became popular, and the fortifications of the period required ironmongery such as gate hinges, locks, and drawbridge chains. The complex shapes and almost invisible welds used in the construction of armour are evidence of the skills the craftsman could command.

From the thirteenth century onwards dies were used by the smith to produce embellishments for the beautiful wrought iron work, examples of which still exist. The hot metal was hammered into a die to produce an impression of a vine leaf or a sheaf of growing corn. During this period the smith's work extended to include the fabrication of cold iron, using files, drills, rivets, and bolts.

The skill of the blacksmith had reached its zenith by the late Middle Ages, and techniques changed little up to the Industrial Revolution. The only important additions to his equipment were the wrought iron anvil and the Oliver. The Oliver was a treadle-operated tilt hammer (fig. 19) which was used from the fourteenth century onwards to augment the sledgehammer. Wrought iron anvils with steel faces were being produced to an average weight of 68 kg by the seventeenth century.

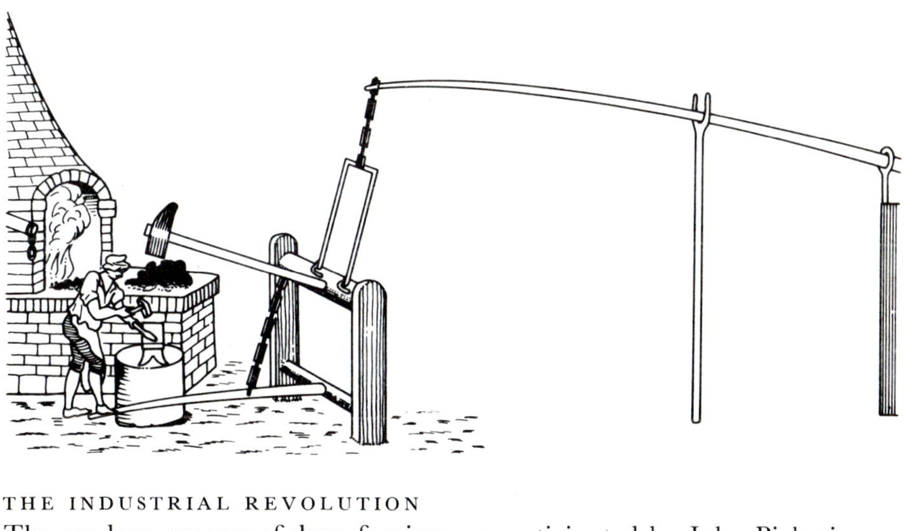

Fig. 19 An Oliver tilt hammer

THE INDUSTRIAL REVOLUTION

The modern process of drop-forging was anticipated by John Pickering, a London toymaker, in 1766. The bottom die was placed on the anvil, the hot metal laid on it, and the top die dropped or hammered with considerable force to produce the object with one blow.

With the scale of working steadily increasing as the Industrial Revolution progressed, the smith could no longer rely solely on the old methods. In 1838 James Nasmyth invented the large steam hammer (fig. 21) which replaced the trip hammer in order to cope with the heavy forging of such objects as the crank shaft of a paddle steamer. The work to be forged was suspended on a

Fig. 20 Water-powered trip hammer

Fig. 21 Nasmyth's steam hammer

heavy chain which allowed the bar to be rotated and moved backwards and forwards while receiving successive blows from the hammer. The model at the Royal Gun Factory, Woolwich, weighed thirty tonnes, and was 10 m high. Nasmyth spoke aptly when he described his invention as 'a thumping child of my own brain'.

The Production of Sheet and Bars

We saw in chapter 1 how malleable iron was shaped into bars in the chafery of the forge. Only two sizes of bars were produced and they were known as 'long' squares and 'short' squares. The long squares weighed about 18 kg and were between 3 m and 4 m in length; short squares were approximately 2.5 cm square in section and weighed about 16 kg. It was impracticable to produce square sections of less than 2 cm in the chafery owing to the loss of heat and the flexibility of the bar.

From these products the blacksmith of the Middle Ages had to work up the size of stock required for any particular job. During the seventeenth century, battery mills for the production of plate or sheet, and rolling and slitting mills for the production of small section bars were established.

Fig. 22 A slitting mill

THE BATTERY MILL

From the seventeenth century onwards these mills handled the production of plate from bar iron. They were equipped with water-powered trip hammers and produced such diverse items as ploughshares, armour plate, and frying pans. Even after the rolling mill had taken over the sheet iron trade, the battery mill was still considered superior for some types of work. In the production of a shovel, for example, it was possible to obtain a gradual taper towards the edge which made for a stronger implement.

ROLLING AND SLITTING

To produce small bars which were less than three-quarters of an inch square it was necessary for small pieces of red hot iron to be passed through rollers before they were slit into bars.

In the late sixteenth and early seventeenth century hand-operated rollers were employed for this purpose, the slitting being accomplished by shears. By 1650, however, water-powered rolling and slitting mills existed. The slitting of the rolled plate was carried out by passing it, still hot, through grooved rollers. The efficiency of these mills was, of course, restricted by the available power, and by the wooden frames used to support the rollers.

Fig. 23 Eighteenth century grooved rolls – Wortley Forge

IMPROVEMENTS AFTER 1784

The introduction of the puddling furnace for the manufacture of wrought iron and the introduction of the steam engine as a source of power rapidly increased the potential of the rolling mill. The bloom from the puddling furnace, which weighed about 250 kg, was passed through a series of rolls known as the roughing mill, and at this stage the product was known as puddled bar. The puddled bars were then put through a further set of rolls known as the finishing train, and by using plain or grooved rolls it was possible to turn out plates, bars, or rails as desired. The use of steam to drive the rolls was the next logical step, and this innovation is credited to the resourceful John Wilkinson.

During the nineteenth century, the technique of rolling progressed rapidly. One of the most important developments was the three-high mill which was perfected by the American, Bernard Louth. It enabled the metal to be passed to and fro without manhandling over the top of the mill.

A continuous strip mill, in which the thickness of the metal was reduced by passing through a series of rolls without interruption, was first operated in this country by George Bedson in 1862. Bedson used his mill for rolling wire, and in the early days the hazards were frightening. As the strip reached speeds of up to 30 m per second any error in the synchronisation of the rolls would send loops of hot metal flying into the rafters of the mill. At the turn of this century, it was found that one way to cope with this faster work was to put the metal on and off the rolls by means of moving tables. In the case of strip, its hurtling process along the rolls would be stopped abruptly at the flying shears where it would be cut into appropriate lengths.

With the introduction of the open-hearth furnace and the Bessemer converter for the production of mild steel, new problems in rolling had to be surmounted. Steel from the furnace was cast into large ingot moulds, the ingot being far larger and heavier (up to 1500 kg) than the bloom from the old puddling furnace. Thus the cogging mill was developed in 1884 to reduce the size of the ingot before passing it along to the strip mill. This was of the two-high reversing type with large rollers which reduced the ingot to a bloom of steel about 12 cm square.

Until the late nineteenth century, all rolling had taken place on hot iron and steel while the metal was still in a plastic state. In the nineties Bernard Louth patented a method of rolling steel when cold. Using this process, steel could be finished with a clean bright surface to high standards of accuracy—within 25 to 50 microns.

Wire Drawing

The method of producing wire by pulling metal through tapered holes in a draw plate was known as early as 2500 B.C. and a number of medieval sources contain pictures of the operation (fig. 24). The draw or wortle plate contained a number of tapered holes, the sizes of which decreased by small degrees. The metal to be drawn was first tapered by hammering, and the tapered end inserted in the largest hole. The operator, seated in a swinging seat, gripped the end of the metal with tongs and pulled it through the hole. Lubrication by tallow or soap was essential, and the face of the plate nearest to the operator was reinforced with steel to minimise wear. Home produced bar iron proved unsuitable for drawing into the fine wire much in demand by the woollen industry for the carding process. The problem was overcome by obtaining the services of smiths from Westphalia, skilled in the production of Osmond iron.

Fig. 24 Wire drawing

This iron was produced from specially selected ores containing a small percentage of manganese, and the fluxing process (see chapter 1) differed somewhat from that used for producing ordinary bar iron.

THE MECHANISATION OF WIRE-DRAWING

Although there are many uses for wire in its original form it is also used for the fabrication of many other articles such as nails, springs, barbed wire, wire ropes, pins and needles. George Bedson of Manchester, it will be remembered, discovered how to produce wire by a continuous process. The hot rolled rod was coiled around a drum, the end of the rod pointed, and passed through the die. The pointed end was then attached to an empty drum which rotated under power, pulling the rod through the die which was lubricated with dry soap.

Wire made from mild steel had to be annealed before it could be used in order to relieve the stresses caused by the drawing operation; whilst carbon steel wire, sometimes referred to as piano wire, had to be both hardened and tempered. This heat treatment was developed as a continuous process by J. Horsefall in 1854, and the wire passed in turn through a furnace, a quenching bath, and finally a tempering bath of molten lead. The process became known in the wire-drawing industry as 'patenting'.

Fig. 25 Mechanised wire drawing

Tube Drawing

Tubes and pipes have an ancient origin. In 2750 B.C. a copper water main was installed in the palace of Sa Hue Ra near the pyramids at Abukir. The pipes were three inches in diameter, and were made by hammering out sheets of copper and rolling them into tubular shapes. The ancient Egyptians went to some trouble to ensure that the pipes were watertight, for the edges were overlapped 25 mm and sealed with non-porous material.

The gun trade and the gas industry provided an impetus to the manufacture of tubes in the eighteenth and nineteenth centuries. The first seamless lead tubes were made by John Wilkinson who cast them around a steel bar.

A method of making tubes by the extrusion process, which entails forcing

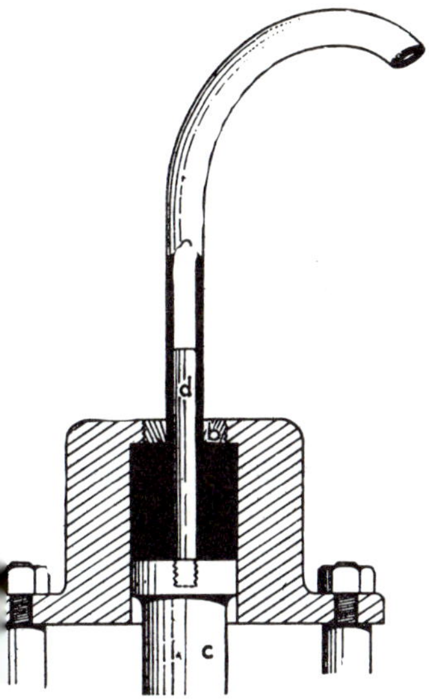

Fig. 26 Burr's extrusion process

metal through a die, was suggested by Bramah as early as 1797. His idea was developed by a Shrewsbury plumber, Thomas Burr, who in 1820 successfully extruded lead pipes using Bramah's hydraulic press to provide the power. A modern process for producing seamless tubes involves punching a hole into a round billet of steel to give a thimblelike formation. The thimble of metal is then placed on a steel mandrel and pushed through a series of dies which reduce the wall thickness and progressively increase the length of the tube.

An alternative method of producing seamless tubing was the Mannesman process (fig. 27), introduced by two German brothers of that name, who opened a mill in Swansea in 1885. This process produced a hollow bloom of metal which could then be rolled into a tube of the required diameter and thickness in the pilger mill. The pilger mill possessed two rolls of irregular semi-circular cross section which exerted a backward thrust on the hollow bloom mounted on a mandrel. A counter thrust from a piston pushed the mandrel holding the bloom through the gap in the rolls thus forging the bloom around the mandrel. The pilger mill took its name from the medieval pilgrims who were supposed to complete their journeys by taking three steps forward and two steps back.

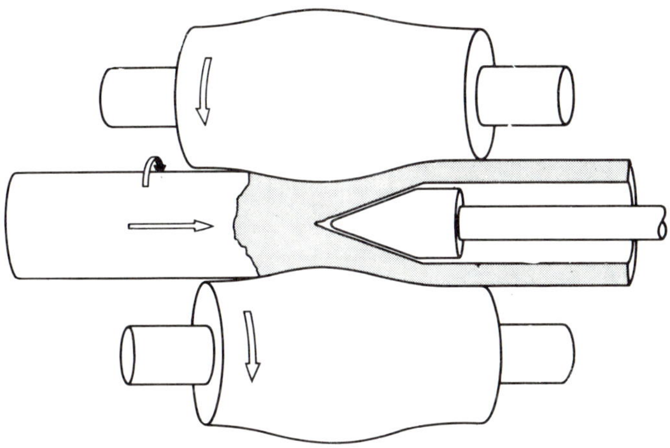

Fig. 27 The Mannesman process

The Coating of Iron and Steel

TINPLATE

The coating of sheet iron with tin was practised in Bohemia, where it was used for decorative articles as early as the thirteenth century. The manufacture of tinplate in this country began shortly after 1720 in the Pontypool area of Monmouthshire. In the early nineteenth century an Englishman, Peter Durand, patented a method of storing food in an iron container coated with tin. This saw the beginning of a demand for tinplate which has increased so rapidly of recent years that its production forms an important part of our economy.

From about 1825 the 'hot dipped' method of tinning was gradually perfected for the coating of iron, and later for coating mild steel. The various techniques of hot dipping were named after the small Welsh mills which pioneered them, giving rise to names like the Melingriffith pot and the Abercarn pot. The basic technique was as follows:

Hot-Dipped Tinplate. Two wrought iron or mild steel bars were first rolled separately, and then together, after which they were twice doubled and re-rolled, giving a pack of eight sheets, each sheet being approximately 250 microns thick. The pack of sheets was then opened and the individual sheets were annealed, pickled to remove scale, and dipped in palm oil. The sheets were dipped twice in the 'tin-pot' and oiled again before passing through rollers which regulated the thickness of the tin coating.

Tinplate manufacture was virtually a Welsh monopoly throughout the nineteenth century. In 1890 the Mackinley Tariff placed a heavy duty on imported tinplate entering the United States. Behind this tariff barrier the Americans were able to take advantage of their rapidly expanding tinned meat and fruit trade to build up a large tinplate industry. By the 1930s the Welsh industry had the ironic experience of having to learn the new electrolytic method from the Americans. The new method, which gave a thinner and more economical coating, enabled the steel to be tinned in the form of a continuous strip produced on the cold strip mill.

Tinplate tends to rust quickly when exposed to the weather, and in 1836 a French chemist, M. Sorel, introduced a method of coating iron sheets by dipping them in molten zinc. The process was called galvanising, taking its name from the famous pioneer in experimental electricity, Galvani. The electrochemical action set up between the zinc and the iron when exposed to moisture resulted in the zinc being corroded and not the iron. Corrugation of iron sheets to increase their load-bearing strength was introduced by R. Walter of Rotherhithe in 1828, and galvanised corrugated sheets have since then been in great demand as roofing material.

Additional Reading

W. H. CHALLONER and A. E. MUSSON, *Industry and Technology*, Vista Books 1963.

T. K. DERRY and T. I. WILLIAMS, *A Short History of Technology*, O.U.P. 1960.

H. R. SCHUBERT, *History of the British Iron and Steel Industry*, Routledge & Kegan Paul 1957.

A. RAISTRICK, *Industrial Archaeology*, Paladin 1973.

Chapter 4
Machine Tools

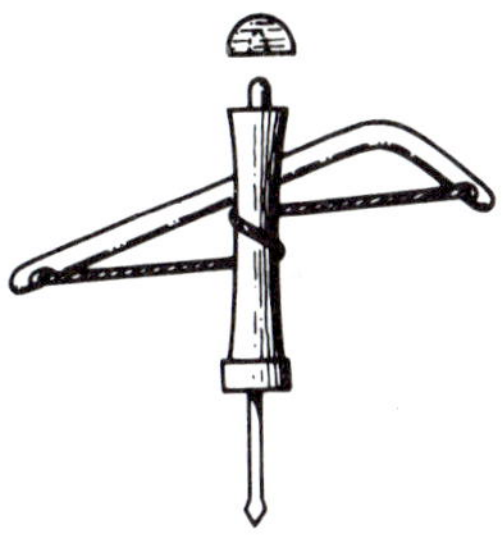

Fig. 28 The bow drill

In engineering, a machine tool is a mechanical device concerned with the removal of metal. The main body of machine tools perform one or more of the following operations: turning, drilling, boring, milling, planing, shaping, and grinding.

Early examples of crude but effective machine tools were the bow drill (fig. 28), which was known to the Egyptians, and the simple bow lathe, which was in use in eastern Mediterranean countries by 900 B.C. In this chapter we shall trace the development of the different machine tools up to the end of the nineteenth century. We shall be describing the work of some famous engineers who made significant contributions in this field, but it should be remembered that very few inventions are entirely original, and that usually there is a history of development over a long period.

The Lathe

THE EARLY LATHES

It is impossible to pinpoint the exact date of the invention of the lathe, but it was certainly known to the ancient Greeks, and it may well have been inspired, in its crudest form, by the potter's wheel.

Its early forms probably resembled the spindle lathe, on which a piece of wood was pivoted between stakes driven into the ground, rotation being provided by pulling to and fro on a string wound round the workpiece. A variation of this drive would be for the turner to provide the power by operating a bow with his left hand.

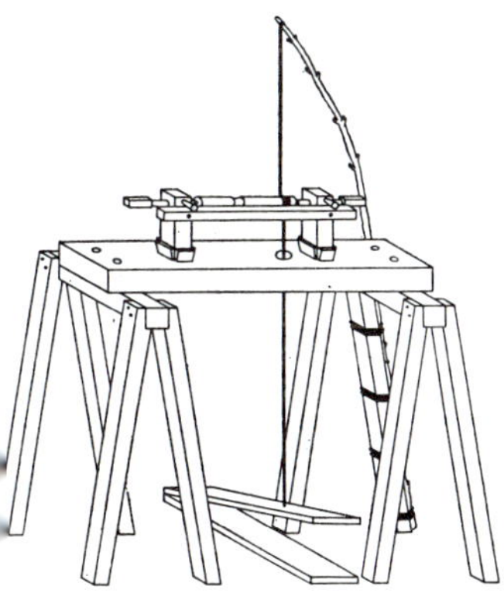

Fig. 29 Pole lathe

An improvement on the bow drive can be seen in the pole lathe (fig. 29) which left the operator free to use both hands. Lathes of this type have been in use from at least the classical Greek period, and it is likely that examples are still to be found today. They were certainly used by the chairmakers, or bodgers, in the Chilterns until quite recently. A big disadvantage of the pole lathe was that it rotated the workpiece backwards and forwards, thus the cutting action was intermittent.

Little improvement was made to the lathe until the late sixteenth century, although earlier Leonardo da Vinci turned his gifted and fertile mind to its development, and left descriptions of a lathe which had fixed and running centres, and was provided with a treadle and crank to give rotation in one direction. Like many other of his ideas it was ahead of its time and did not influence contemporary machine builders.

In 1569 Besson published details of a lathe (fig. 30) which was capable of producing screws, using a leadscrew which carried forward a rigidly mounted tool. The diagram shows a wooden tapered screw being turned on this lathe; it was manually operated, and must have been slow and extremely clumsy.

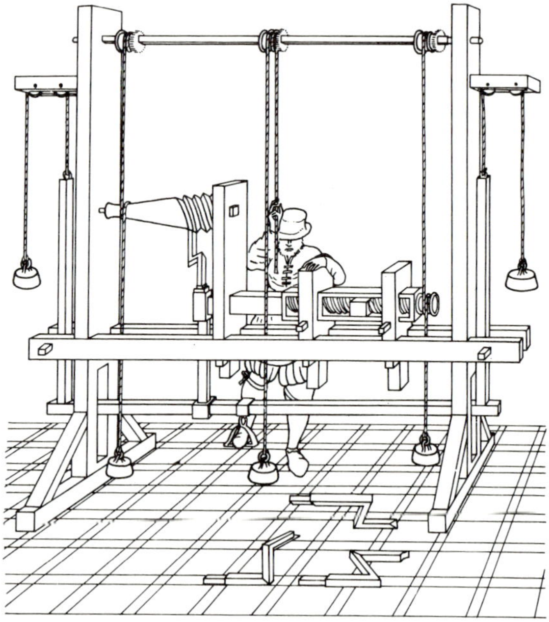

Fig. 30 The Besson lathe 1569

THE MANDREL LATHE

The first big improvement on the pole lathe came in the late sixteenth century with the introduction of the mandrel lathe. The drive was removed from the workpiece, which had been a limiting factor to the scope of the pole lathe, and applied to a mandrel (or spindle) connected to a live centre. The workpiece was held between the live (or running) centre and a fixed centre. This modification to the drive also made possible the use of a primitive type of

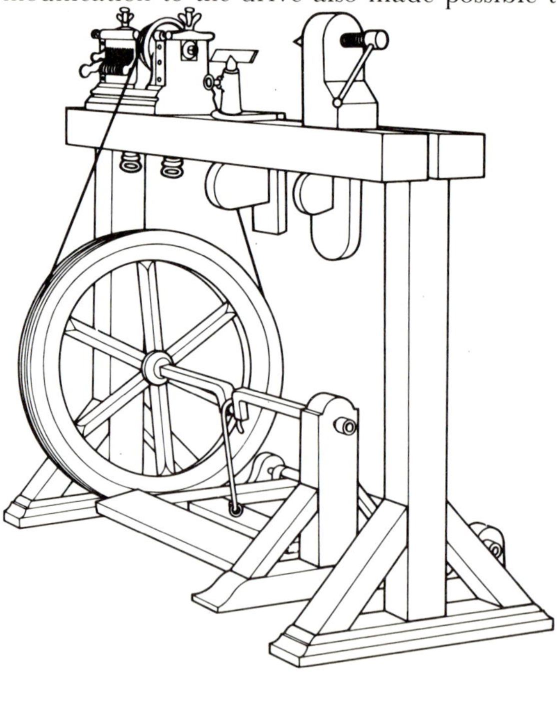

Fig. 31 Mandrel lathe

chuck for holding any work unsuitable for turning between centres. By 1700 the mandrel was being made of iron running in bearings of cast tin or lead, and fig. 31 shows the lathe when it had reached its most developed form in 1785. This lathe was capable of generating screwthreads, and the action of the mandrel can be explained by referring to the plan of the headstock (fig. 32).

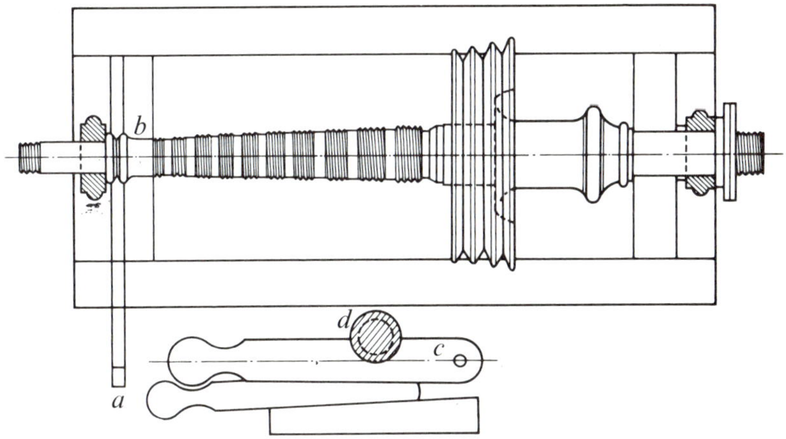

Fig. 32 Mandrel lathe headstock

For plain turning a master key (*a*) was wedged into the large groove (*b*) at the left of the mandrel, allowing rotary motion only. To cut a screw thread the master key was depressed, and a guide key (*c*) wedged into the appropriate guide screw (*d*) on the mandrel. When the mandrel was rotated the action of the guide key in the guide screw carried the mandrel and workpiece axially across a cutting tool held stationary on the tool rest by the operator. In this way a replica of the guide screw was formed on the workpiece. The length of screw being cut was not limited by the length of the guide screw as the turner could take up a new position with the tool, and repeat the operation.

The range of the mandrel lathe was limited by the number of guide screws cut on the mandrel, and the accuracy of the work produced was affected by the wooden construction of the lathe bed and the headstock.

THE LEADSCREW LATHE

The clockmakers of the early eighteenth century made considerable headway in improving the accuracy of the lathe, probably because they were the only metalworkers whose product demanded a degree of precision. A French watchmaker's lathe of about 1740 was of all-metal construction, and had a leadscrew connected to the workpiece by gears. The slide rest, a device for carrying the rigidly mounted tool across the workpiece, was introduced in France about 1760.

The search for a quicker and more accurate method of making screws accelerated the development of the leadscrew lathe at the dawn of the Industrial Revolution. That the basic principle of the leadscrew had been understood for some considerable time is clear from the Besson drawing, and it would be as well to clarify the principle at this stage. A screwthread may be generated on a rod carried in a lathe if it can be connected to a long screwthread (leadscrew) by means of gearing. The leadscrew must propel a slide rest, holding a tool ground to the profile of the thread, axially along the workpiece as it rotates. The number of threads per inch (t.p.i.) cut on the work-

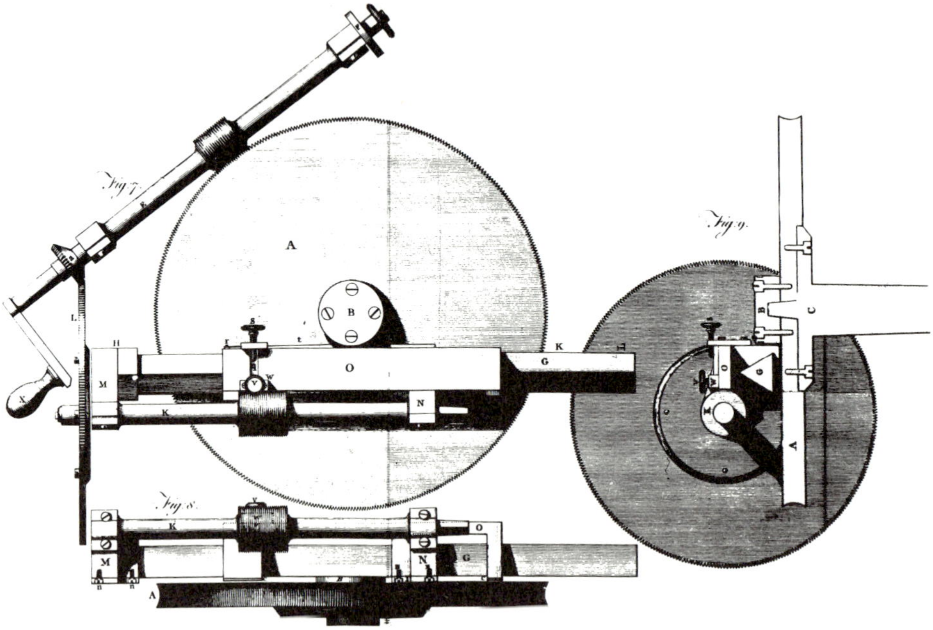

Fig. 33 Ramsden's lathe showing leadscrew and gearing circa 1770

piece depends on the ratio of the gears connecting the mandrel to the leadscrew and on the t.p.i. of the leadscrew itself.

The first practical screwcutting machine embodying these features is ascribed to Jesse Ramsden, an instrument maker, and was introduced about 1770. The leadscrew and gearing can be seen in fig. 33. The importance of this lathe is difficult to assess but it is quite possible that its construction influenced later work on similar machines. Two Frenchmen, Jacques de Vaucanson and François Senot, working independently, built a form of screwcutting lathe at the end of the eighteenth century and an American, David Wilkinson is reputed to have built one about the same time.

HENRY MAUDSLAY (1771-1831)

It was left to Henry Maudslay to build the first screwcutting lathe of commercial value in the closing years of the eighteenth century. The lathe (fig. 34) had a bed comprising two bars of triangular cross-section which carried

Fig. 34 Maudslay's first screwcutting lathe

the slide rest. The headstock carried a live spindle (mandrel) which was connected to the leadscrew by a pair of gears. The lathe was provided with several interchangeable leadscrews of differing t.p.i., so that screws of different pitches could be cut. The slide rest was incorporated for the first time as a practical feature, and for many years this part was referred to as 'Maudslay's Carriage'. This lathe can be considered as the prototype of the modern machine.

Fig. 35 Maudslay's small screwcutting lathe circa 1800

Maudslay improved on his design so rapidly that by 1800 he was using a lathe (fig. 35) which had a substantial cast iron bed, and he had conceived the idea of using change wheels to avoid interchanging the leadscrews. This lathe had, in fact, twenty-eight change wheels, and the sample screwthreads ranging from 16 to 100 t.p.i. can be seen in the rack in the front of the photograph of the lathe.

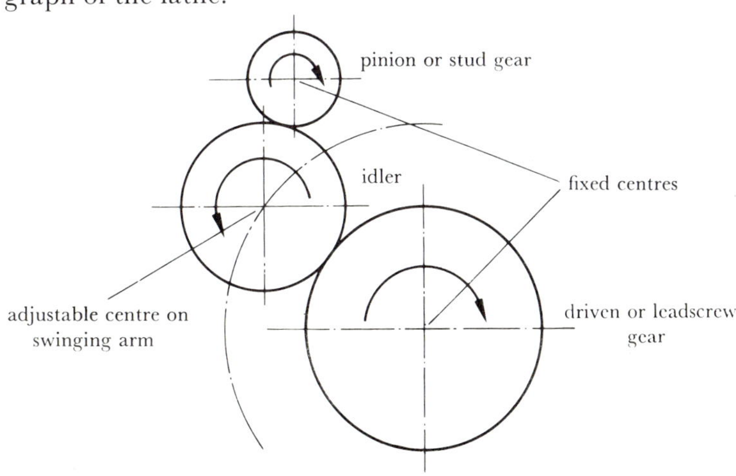

Fig. 36 The principle of change wheels

The theory of change wheels can be explained with reference to fig. 36. For any particular screwthread the number of teeth on the gear wheel can be arrived at as follows:

$$\frac{\text{t.p.i. of leadscrew}}{\text{t.p.i. to be cut}} = \frac{\text{no. of teeth on pinion}}{\text{no. of teeth on driven gear}}$$

thus, if a screwthread of 20 t.p.i. is to be cut, and the leadscrew has 8 t.p.i., a pinion having eight teeth and a driven gear having twenty teeth will be required, or a convenient multiple of that ratio, e.g. 16/40, 32/80, etc.

Metric threads are designated by their pitch, i.e. the axial distance between the crests of adjacent threads. The change wheel formula becomes:

$$\frac{\text{Pitch of thread to be cut}}{\text{Pitch of leadscrew}} = \frac{\text{no. of teeth on pinion}}{\text{no. of teeth on driven gear}}$$

Therefore to cut a thread of 2 mm pitch on a lathe with a leadscrew of 6 mm pitch would require gear wheels with teeth in the ratio of 2:6, e.g. a 20-tooth pinion and a 60-tooth driven gear.

To enable the pinion and driven gear to be changed at will, Maudslay introduced the method of mounting an idler gear on a swinging arm which would mesh with gears of varying diameter mounted with a fixed distance between centres. He was also concerned that the guiding surfaces of the lathe should be finished to high standards of accuracy, and he introduced the use of hand-scraped surface plates to test these flat surfaces.

The startling advances made by Maudslay are more readily appreciated when we consider that at the time he introduced his lathes the wooden pole lathe was still in widespread use.

ROBERTS (1789-1864), FOX (1789-1859), AND WHITWORTH (1803-1887)

The next name of importance in connection with the development of the lathe is that of Richard Roberts, an ingenious engineer who made a remarkable contribution to production and textile engineering. His lathe of 1817 introduced the four-step pulley and back gear which gave a range of eight spindle speeds (fig. 37). He also introduced the self-acting traverse which enabled the tool to pass along the workpiece automatically. This was achieved by a long screw, additional to the leadscrew, driven from the main spindle by a type of bevel gearing.

A contemporary of Roberts was James Fox of Derby, the only significant toolmaker of the era who had not worked under Maudslay. Primarily a manufacturer of textile machines, he nevertheless made some important contributions to the development of the lathe. He introduced the separate keyed drive shaft to operate a pinion which then engaged with a rack on the bed to move the carriage when plain turning. This meant that the leadscrew need only be used for screwcutting and that consequently its accurate life was prolonged. Fox was also responsible for the inverted vee and flat guideways on the bed which increased the accuracy of the machine.

Joseph Whitworth, one of a number of Maudslay's pupils to continue his work, set up as a machine tool manufacturer in 1833. He produced a lathe with a rigid boxlike bed, and an automatic cross-feed; he also introduced the

Fig. 37 Robert's back-geared lathe 1817

Fig. 38 A Whitworth lathe 1843

method of engaging the leadscrew with a halfnut. Whitworth is probably best remembered for his work on the standardisation of screwthreads, but his machine tools were world-famous in the middle of the nineteenth century.

THE AMERICAN CONTRIBUTION

To study the continuing development of the lathe we must now look at the work of the American engineers, because from the middle of the nineteenth century onwards many of the developments in machine tools took place in the U.S.A. Although we have not room here to study the reasons for this in detail, it had much to do with the rapid expansion of their economy and the readiness of the Americans to use machines to offset their lack of skilled manpower. Much of the progress took place in the gun-making industry where they developed the use of interchangeable parts produced by specially designed machines operated by semiskilled or unskilled labour. Modifications were made to the lathe to achieve this end, and Stephen Fitch of Middlefield, Connecticut, introduced the turret lathe in 1848. The turret held eight different tools, and it was only necessary to rotate the turret to bring the next tool into its cutting position.

By 1873 Christopher Spencer had perfected a fully automatic lathe (fig. 39). Two large drums, referred to as 'brain wheels', carried adjustable cams which were mounted beneath the lathe bed. One of these drums controlled the operation of the collet chuck while the other drum governed the sequence of machining operations. Spencer's next contribution to increased productivity was to design an automatic lathe with three spindles. This enabled three different machining operations to be carried out simultaneously in the production of small screws. The multi-spindle lathe was further developed by Reinhold Hakewessel who built a four-spindle automatic lathe which was selling on the open market at the turn of the century.

Fig. 39 Spencer's automatic lathe c. 1873

Towards the end of the century the discovery of high-speed steel, by the American F. W. Taylor and his associates, greatly increased the productivity of the lathe, as indeed it did that of other machine tools. In consquence, lathes were made heavier and more rigid to withstand the increased cutting performance of the tools. Taylor also showed that economical machining could be obtained by greater feed and drive power. To this end he carefully worked out exact formulae for feed speeds and tool pressures.

Only two more important improvements need be mentioned to arrive at the modern type of centre lathe. The first of these was the introduction of the Norton quick-change gear box in the early nineties, which enabled the appropriate gear train for any particular screwthread to be selected by means of a lever. The second was the fully geared headstock, which appeared about the turn of the century, with clutch operated drive from a built-in constant-speed electric motor enabling a wide range of spindle speeds to be selected quickly and easily.

The Milling Machine

Milling is the removal of metal from a workpiece by passing it beneath or across a rotary cutting tool.

The concept of the milling machine was understood by French clockmakers of the eighteenth century, and Jacques de Vaucanson used such a machine for cutting gear wheels for clocks. However, the development of the miller as a viable commercial machine tool stands to the credit of American inventors, and its introduction was influenced by the economic considerations mentioned previously.

THE FIRST MACHINE

Fig. 40 Eli Whitney's milling machine circa 1820

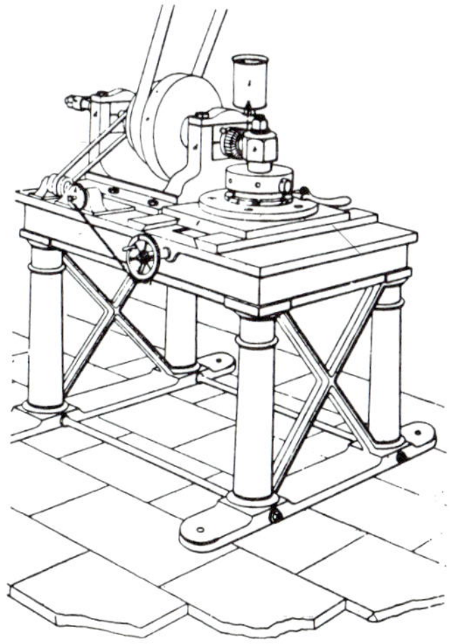

Fig. 41 Nasmyth's miller 1830

The earliest machine which can be recognised as fulfilling our definition has been credited to Eli Whitney and this machine was in use about 1820. Whitney, an American, who was already renowned through his invention of the cotton gin, turned his attention to the mass production of rifles when his patents on the cotton gin failed to bring him the reward to which he was entitled. His machine (fig. 40) incorporated an automatic feed for the worktable (missing from the illustration) operated by a screw and nut; the automatic feed could be disengaged to allow hand feeding by dropping the handle carrying the worm. The use of the box principle of construction for the base, although made of wood on his machine, was an essential requirement for rigidity. Whitney also made use of gibs for holding down the table and taking up wear, but the absence of a provision for vertical adjustment of the workpiece limited the machine's usefulness.

There is evidence that another American, Robert Johnson, introduced a milling machine about the same time as Whitney. It appears to have been inferior in many respects to Whitney's machine, lacking, among other features, the automatic feed.

In 1830 James Nasmyth made the only significant British contribution in this field. His milling machine (fig. 41), designed especially to machine the flats of hexagonal collar nuts, was constructed at the time of his apprenticeship to Maudslay. It was robustly constructed with a metal frame. Although no means of vertical adjustment was provided and the feed mechanism was weak and intermittent in action, it did have an indexing arrangement to position accurately the hexagonal faces. A can was fitted to the machine to supply a continuous flow of water directly onto the cutting tool—the earliest evidence we have of the use of coolant on machine tool cutters.

FURTHER DEVELOPMENT IN THE UNITED STATES

In America the development of the milling machine continued with the introduction of a machine by the firm of Gay and Silver about 1835, which incorporated two new features: (*a*) provision for adjusting the height of the cutter spindle, and (*b*) an overarm for supporting the spindle at its outer end. In 1885 the Lincoln miller (fig. 42) was introduced by the firm which gave it its name. It was designed by E. K. Root assisted by F. A. Pratt, and was inspired by the earlier work of an engineer of considerable talent, F. W. Howe.

The Lincoln miller was the first production milling machine, and many

Fig. 42 Lincoln miller 1855

thousands were built and sold throughout the world. Although these machines were of robust construction and featured a number of refinements, the problem of vertical adjustment of the workpiece remained unsolved, and height adjustment had still to be made through the cutter spindle.

An important refinement to the design of the milling machine was probably inspired by the problem of machining the helical flutes in twist drills. Twist drills rapidly replaced the old flat drills in the 1850s, and the filing of the flutes by hand called for much skill and patience.

Fig. 43 Universal miller 1862

At the suggestion of Howe, now employed by the firm of Brown and Sharp, Joseph Brown designed a milling machine to undertake the task. Such was the genius of Brown that not only did he succeed in solving the problem, but in addition he produced a machine which embodied all the basic features of the milling machine as we know it today. The features of Brown's machine, which entitled it to be called a Universal Miller, were:

(*a*) Vertical adjustment of workpiece using a screw-operated knee which could slide up and down guides on the main frame.

(*b*) Spiral head with the arbor geared to the feed mechanism to enable spirals to be cut and indexed.

(*c*) Provision for machining tapered spirals (e.g. tapered reamers).

MILLING CUTTERS

The milling cutters used by the French clockmakers and the early American machinists were in substance rotary files, having teeth formed by careful work with the hammer and chisel. A Swiss engineer, John Bodmer, working in Manchester was using a formed milling cutter to produce metal gears during the 1830s. He also pioneered what is now called gear hobbing and examples of his gear cutters are still preserved in the Science Museum in London.

Fig. 44 Phoenix milling cutter

The first progress towards the type of cutter we are now familiar with can be seen in the photograph of one manufactured at the Phoenix Iron Works about 1850 (fig. 44). Though the cutter would probably have taken a substantial chip, the labour and time involved in sharpening would have been considerable.

In 1864 Joseph Brown patented a formed cutter (fig. 45) which could be ground on its face without changing its shape. It had eight widely-spaced teeth, accurately machined to the involute form, and was designed specifically for cutting gear teeth. Brown adopted the involute after the superiority of this form over the favoured epicycloidal tooth had been demonstrated by the Cambridge mathematician, Robert Willis.

By 1900 the spiral cutter and the inserted tooth cutter were in use, and special machines for grinding cutters were available. The mass production of the bicycle and the sewing machine depended to a large extent on the versatility of the milling machine.

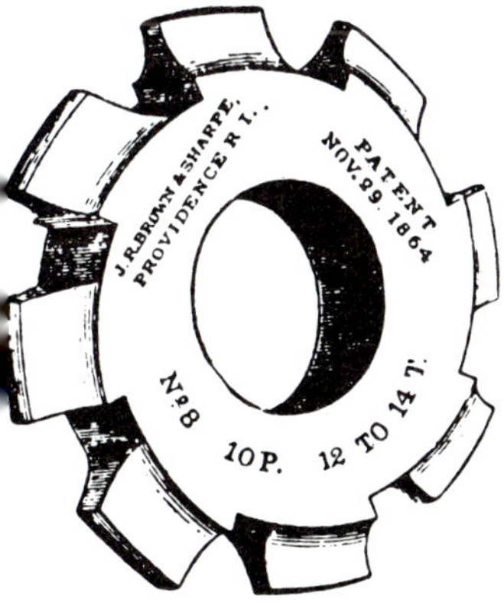

Fig. 45 Joseph Brown's formed gear cutter 1864

Drilling and Boring

THE DRILLING MACHINE

As we stated at the beginning of the chapter, the bow drill was known from the earliest times, but there is little evidence of any substantial improvement before the middle of the eighteenth century. The earliest surviving drill, a horizontal machine, was made by Vaucanson, and a type of pillar drill was in use about 1750. Pillar, column, and radial drilling machines were in use by the middle of the nineteenth century and their efficiency was increased when the old flat or spear drill was replaced by the twist drill. An early type of radial drill (fig. 46) had a column which could be raised or lowered by means of a tommy bar, and the drive was provided through bevel gears (note the spear drill).

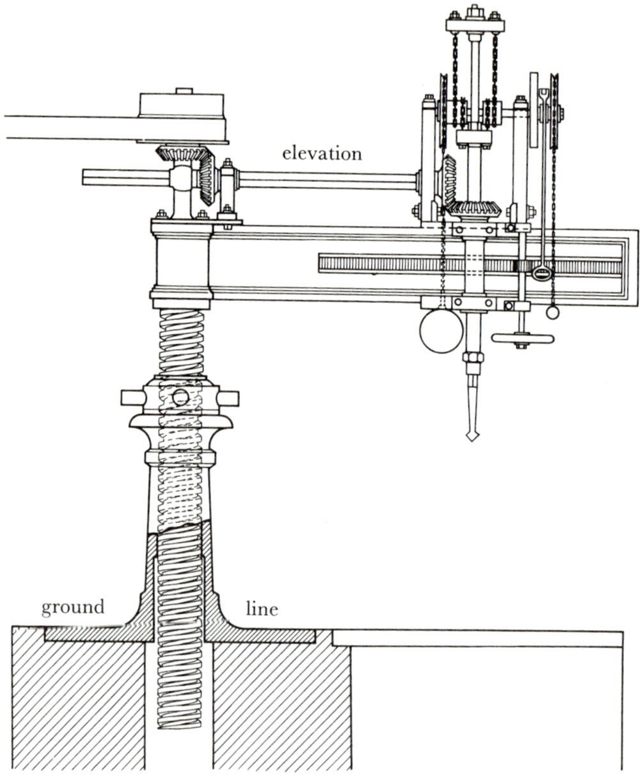

Fig. 46 Radial drilling machine

The first multispindle drilling machine was introduced in the 1850s; the one used for drilling the plates for the Hungerford bridge had eighty spindles, and incorporated hydraulic feed.

THE BORING MACHINE

The boring machine, developed for the boring of cast gun barrels, was in fact the first machine tool available for heavy work in metal. It was a horizontal machine, and as the only support for the boring bar was at the driving end, the cutter tended to follow the cast hole, and accuracy was virtually impossible. James Watt recognised the imperfections of these machines when he first attempted to build a full scale prototype of his steam engine. Such was the inaccuracy that the diameter of the bored cylinder varied three-eighths of an inch between ends, and a steam-tight seal, on which the success of his engine depended, was unobtainable. It is doubtful whether Watt's engine would have become a commercial proposition but for the inventive genius of John Wilkinson, the ironmaster, who, in 1774, patented a boring machine (fig. 47) capable of machining to the accuracy required for Watt's cylinder. This was the first machine tool capable of the accurate removal of metal on large, heavy work (Wilkinson designed it to improve the boring of the cannon cast at his works) and it is worth closer examination.

The cylinder to be bored was rigidly mounted, and a hollow boring bar, slotted along its length, passed centrally through it supported by bearings at either end. A disc which carried the cutting tools was mounted on the boring bar and was attached, through the slot, to a rod inside the bar. This rod did not rotate (see inset) and could be fed along the bar by means of a rack and pinion.

By 1800 the vertical boring mill had been introduced and was increasingly favoured for heavy work. It was, in fact, a vertical lathe, the workpiece being bolted onto a large horizontal face plate which rotated while the boring bar was fixed.

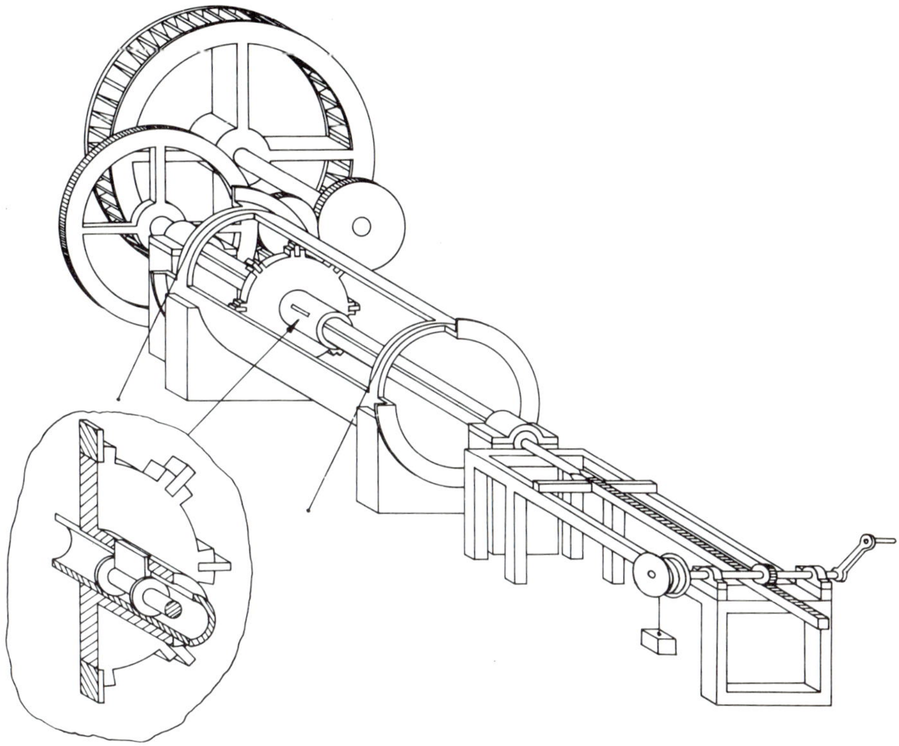

Fig. 47 Wilkinson's boring machine

Grinding

From the earliest times natural abrasive stones have been used for sharpening weapons and tools, and water-powered grindstones were being used by the sixteenth century. The development of the grinding machine was largely the work of American engineers, although the Frenchman, Réhé, was using a tool and cutter grinder as early as 1783.

In 1834 Wheaton of Providence, Rhode Island, introduced a grinding machine which contained many features of an advanced nature, including provision for grinding tapers. Nasmyth produced a grinder in 1845 which featured two cast iron wheels, 2 m in diameter, on which were mounted abrasive stones. The significance of this machine, with its automatic feed, has perhaps been underestimated.

The introduction of the artificial stone towards the middle of the century hastened the development of the grinder. These stones had grits of iron oxide or aluminium oxide and silica, bonded with clay or rubber. The improved universal grinding machine introduced by Brown and Sharpe in 1876 contained most of the features to be found in a modern machine, including provision for form grinding. The discovery of carborundum (artificial silicon carbide) by E. A. Acheson in 1891 improved the durability and cutting power of grinding wheels. Thus, by 1900 the grinding machine was well established as a standard method of removing metal as well as being used for sharpening purposes.

Planing and Shaping

So far we have discussed machine tools in which the cutting action is derived from rotary motion. There remains another class of machine tools whose cutting action is based on a reciprocating motion of the workpiece or the tool. These actions are obviously derived from hand tools such as the file and the plane.

These machine tools fall into two main categories:

(*a*) the planer, in which the workpiece moves backwards and forwards under a stationary cutting tool,

(*b*) the shaper, which has a reciprocating cutting tool with the workpiece held stationary.

The slotter, in which the tool reciprocates vertically, is a derivation of the shaper.

THE PLANER

Joseph Bramah patented a wood planer in 1802. This planer, which carried twenty-eight cutting tools, was in use at Woolwich Arsenal for over fifty years. He also used a planer to produce the metal components of his famous locks, but little is known about it.

Fig. 48 Whitworth's planing machine 1842

A number of British engineers appear to have produced planing machines in the early years of the nineteenth century, and it is difficult to ascribe the invention to one individual. Three names are worth noting in this connection however—Roberts, Clement, and Murray.

Matthew Murray, a rival of Boulton and Watt in the production of steam engines, was using a planer in his Leeds factory by 1814.

Richard Roberts, whose name has appeared in connection with the lathe, produced his planer in 1817. It had many notable features, including vertical and horizontal feed and a hinged clamp (clapper box) to allow the tool to lift on the return stroke. The table measured 132 cm long by 28 cm wide.

Joseph Clement, who had worked with both Bramah and Maudslay, built his first planer in 1820 to machine the triangular slide bars of lathes. Some years later he built his 'great planer' which incorporated two cutting tools, one for use on the forward stroke and one for the return stroke. It could accommodate work up to 182 cm square, and for many years was the only one of its size.

Whitworth considerably improved the design of the planer. The one he introduced in 1842 (fig. 48) was power driven, and the tool holder reversed at the end of each stroke, giving cutting action in both directions.

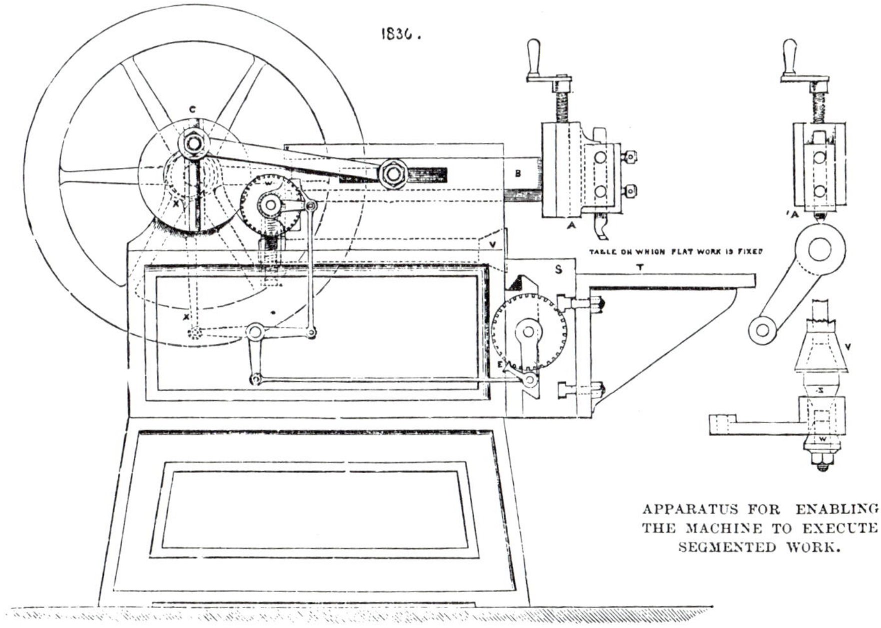

Fig. 49 Nasmyth's shaper 1836

THE SHAPING MACHINE

James Nasmyth invented the shaping machine in 1836. The machine (fig. 49) was hand powered, but included many features to be found in a modern shaper. Movement of the worktable along the dovetail slides of the saddle was automatic, the drive being provided by a pawl and ratchet mechanism. The tool was housed in a clapper box with a head slide to provide vertical adjustment, and the workpiece was held rigidly in a substantial vice. Shortly after the introduction of the shaper, Whitworth designed the mechanism to reduce the time taken by the return stroke.

The Age of Machine Tools

In 1802 Maudslay commenced the construction of forty-four special machines to mechanise the manufacture of wooden pulley blocks for the Navy. These machines were inspired by Sir Samuel Bentham, and were designed by Marc Isambard Brunel. By 1808 they were in full operation, and enabled ten unskilled workmen to produce some 130,000 pulley blocks a year, a task which had previously occupied 110 skilled workmen. This was a striking demonstration of the economy of machine tools, but its potentialities were not appreciated in this country.

It was left to the Americans to develop the use of machinery to produce standardised interchangeable parts, and among the pioneers in this work were Eli Whitney, Simeon North, and Samuel Colt. The 'American System' as it was called, gave the United States a worldwide reputation for the manufacture of firearms, sewing machines, and agricultural machinery in the middle of the nineteenth century.

Sir Joseph Whitworth was a member of a commission which went to America to investigate these matters, and was full of praise for what he saw. As a result, the American firm of Robbins and Lawrence was asked to build and install machinery for the manufacture of the Enfield rifle on the lines of the 'System'.

The introduction of the assembly line technique for the manufacture of motor cars by Henry Ford was a brilliant but logical extension for the use of machine tools and fully merited the term 'mass production'. On the debit side mass production meant the acceptance of monotonous and impersonal working conditions and the loss of craft status.

Additional Reading

A. E. BURSTALL, *A History of Mechanical Engineering*, Faber 1963.
J. W. ROE, *English and American Toolbuilders*, Yale U.P. 1916.
L. T. C. ROLT, *Tools for the Job*, Batsford 1965.
R. S. WOODBURY, *History of the Milling Machine*, M.I.T. Press 1960.

Chapter 5
Sources of Power 1

The history of man's progress from the primitive state has depended upon his ability to harness the available sources of natural energy. In the beginning it was the energy of his own body. At a later stage, he harnessed the energy of animals and at a much later stage he discovered how to harness the natural forces of wind and water. Power requires machines both to generate it and to use it. A machine which generates power is called a prime mover. Prime movers convert the natural forces of the body, wind, running water, fuels, nuclear fission and chemical action into mechanical energy. Thus muscles, windmills, water wheels, petrol, steam and diesel engines, nuclear power plants and electric motors driven by batteries are all prime movers. In this chapter and the next we shall be considering the historical development of prime movers, the sources of power.

Man and His Muscles

Over the ages man developed simple machines to multiply the force he could exert with his muscles. It was this ability which established his superiority over the other primates. In probable order of discovery the six basic machines that man developed were the lever, the inclined plane, the wedge, the wheel and axle, the pulley and the screw. All of these had been discovered by 200 B.C. It is an interesting fact that all machines, no matter how modern and complex, can be resolved into a combination of some or all of these six basic machines.

The wedge and the lever were used in pre-historic times. The wedge was particularly useful for splitting wood and large rocks. To split rocks, wooden wedges were hammered into cracks and were then soaked with water. As the wood swelled the cracks were forced apart by the pressure exerted. The blue stones used to build part of Stonehenge were hewn from the Prescelly mountains in this fashion.

An early example of the lever was the shaduf, used in Egypt by 2500 B.C. The illustration comes from an Egyptian tomb and is dated about 1500 B.C. (fig. 50). The shaduf consists of a long lever with a water container at the shorter end; the counterbalance at the other end of the lever was made of mud, as was the supporting pillar or fulcrum. The shaduf was used mainly for irrigation, but it was also employed for other purposes as it enabled a man to lift many times his own weight simply by applying pressure on the counterbalance end.

Fig. 50 Egyptian shaduf circa 1500 B.C.

The theory of the lever was first described by the Greek Archimedes, who will be remembered for his work on displacement and for his water-raising screw.

One of the earliest mechanical problems that had to be solved was the ability to convert reciprocal motion into rotary motion and vice versa. The bow, which was first used as a weapon, was also used from very early times for making fire and for drilling holes (fig. 28). The bow holds particular interest for the engineer in that it was the first machine capable of storing power. The opposite problem of converting rotary motion into reciprocating (back and forward) motion was not solved until much later. The various solutions involved the use of the crank or levers and cams.

The Wheel

The discovery of the wheel was of vital importance to human progress. Its first application in the field of transport was most likely arrived at by a progression from the sledge to the roller, and then to the simplest form of solid wooden wheel built up in segments. Its earliest appearance as a machine was the potter's wheel when it had the effect of greatly reducing the muscular effort on the part of the potter. At the same time it made possible the achievement of a perfectly symmetrical form.

The use of stone wheels for grinding grain can be traced as early as 1200 B.C., and the rotary quern or hand-mill used by the Romans incorporated what is possibly the earliest use of the crank (fig. 51).

Fig. 51 Roman quern

Water

THE NORSE OR GREEK MILL

The first machine for harnessing the power of running water was the Norse or Greek water-mill. In spite of its name this mill was probably used as early as 1000 B.C. and originated in the hilly districts of the eastern Mediterranean countries. It was in widespread use throughout Europe during the Middle Ages, and continued in use in some remote parts into this century.

Fig. 52 Primitive Greek mill

The Greek mill (fig. 52) comprised a series of inclined paddles set around a central axis. The paddle was immersed in a swiftly flowing stream, and the axle rising out of the water passed through the lower horizontal mill stone and was secured to the upper one which it rotated. The mill stone revolved at the same rate as the paddle since there was no intermediate gearing.

THE VITRUVIAN WATER-MILL

This type of mill with vertical wheel and horizontal axle was introduced by Vitruvius, a Roman, in the first century B.C. It was probably inspired by the Persian wheel, in common use at that time. This was a vertical wheel with scoops arranged around its rim for lifting water from rivers. The Vitruvian water-mill was arranged so that the lower paddles were immersed in the stream and were moved by the flow of water. The horizontal rotation of the axle was converted into vertical rotation through a lantern gear. This consisted of wooden pegs fixed around the circumference of a vertical wheel and meshing with spaced pegs, resembling a lantern, fixed to the vertical shaft. This is probably the earliest example of the use of gearing to transmit power. It had the added advantage of allowing the speed of the mill stone relative to the water-wheel to be varied by changing the ratio of the gears. Mills of the Vitruvian type with the water passing under the wheel are known as undershot water-mills, and such mills, in operation by the fourth century A.D., were capable of grinding at the rate of four tonnes an hour. These water-mills were not used extensively by the Romans before 300 B.C. despite their superiority over hand-operated mills. This may have been due to the cheap and extensive slave labour available to them before that date.

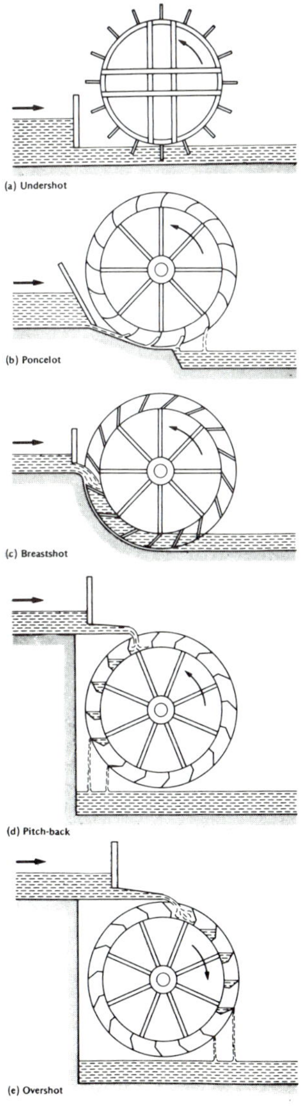

Fig. 53 Types of waterwheel

THE DEVELOPMENT OF THE WATER-MILL

The water wheel was the major power source throughout the Middle Ages. The efficiency of the wheel was increased by the development of alternative methods of delivering the water to the wheel. The different methods are shown in fig. 53. The efficiency of the undershot wheel was greatly increased by the French engineer T. V. Poncelot who introduced an adjustable inclined sluice in 1824.

Water-mills provided power not only for grinding but for many other industrial processes such as saw mills, ore crushers, and trip hammers. They were also used to operate cranes and pumps, and the mechanical skills acquired by the millwright in adapting the machinery to fulfil the many different tasks, were to form the basis of engineering technology in later centuries. An adequate water supply to operate the wheels governed the location of industry, and it was rarely possible to operate waterwheels throughout the year. To improve this situation dams to hold water supplies and gates to control the flow of water through the mill-races were constructed.

From 1582 until 1820 water-wheels were used at the London Bridge waterworks to operate force pumps which raised water from the Thames. The wheels were reversible and, being operated by the tidal flow, were in almost constant operation.

The eighteenth century saw the appearance of the steam engine as an alternative and more reliable power source, but the water-mill was well established and was not to be seriously challenged as a source of *rotary* power until the nineteenth century. Indeed, the early steam engine was used in some cases to pump water into the mill dam so that operation of the wheel could be independent of natural supply.

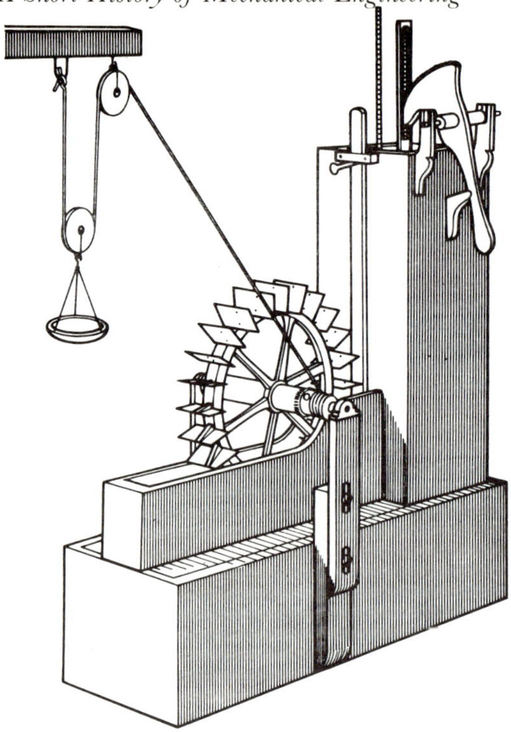

Fig. 54 Smeaton's experimental water wheel

Smeaton, whom we shall meet later in this chapter, conducted a scientific study of water-wheels in the middle of the eighteenth century. He used working models for his experiments, and calculated the efficiency of the wheel under various conditions. He found the efficiency of an overshot wheel to be in the region of 60 per cent compared with the 23 per cent of a similar undershot wheel. Using the knowledge thus acquired, he undertook the design of several water-mills, including a new installation for London Bridge with a wheel 9.75 m in diameter and 4.7 m wide. Smeaton pioneered the use of iron in the construction of water-mills, a practice which led to the gradual replacement of wood which had been used for machine parts since pre-Roman times.

Fig. 55 Foster Beck Mill, near Pately Bridge

WATER TURBINES

The first radical changes in the design of water-wheels since the Vitruvian mill were introduced in the 1820s. Benoit Fourneyron perfected what became known as an outward-flow water turbine in 1827 (fig. 56). The moving portion of the turbine carried curved blades attached to a vertical axis and revolved around fixed guide vanes through which the water passed before meeting the blades of the runner. This arrangement ensured that the water met the moving blades with minimum shock, a necessary condition for efficiency. Early turbines of this type achieved speeds of over 2000 r.p.m. from a water head of 106 m.

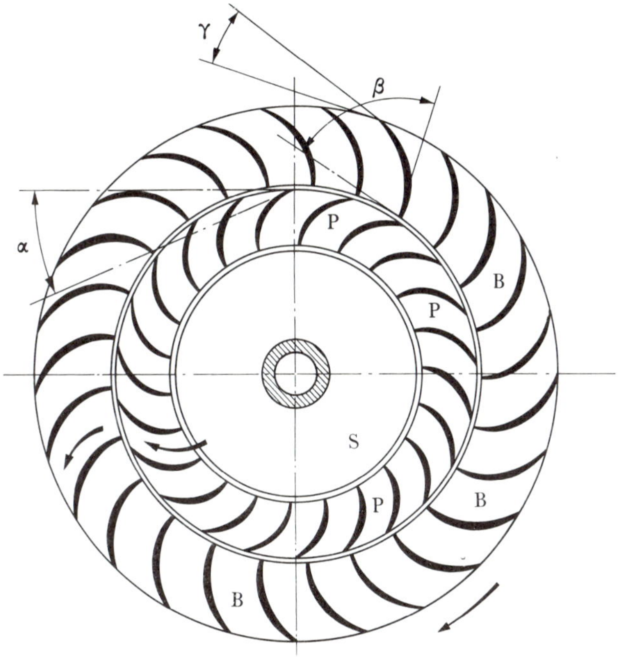

Fig. 56 Outward flow turbine

In 1828 an American, J. B. Francis, introduced the inward-flow radial turbine. This turbine also had a vertical axle, and water entered the enclosed runner radially. The Jonval turbine which appeared in 1843 had an axial flow and in 1852 James Thomson introduced the vortex inward-flow turbine which incorporated pivoted guide vanes to control the performance of the turbine on part load.

With the invention of the electric dynamo the water turbine attained new importance as it was eminently suited to provide the high speeds required by the dynamo. The first hydroelectric installation was in use at the Niagara Falls by 1895, and installations followed in other countries, particularly Norway and Sweden, where natural conditions provided suitable heads of water.

Finally, mention must be made of the Pelton wheel which was developed in the goldfields of Alaska in the 1870s. This wheel (fig. 57) had hemispherical cups placed at intervals around the circumference. These cups or buckets had an internal partition at which the jet of water was directed. This partition caused the water to flow tangentially over the surface of the bucket and greatly reduced the shock which had detracted from the efficiency of earlier wheels in which the jet struck the bottom of the bucket. Pelton wheels have developed 375 kw from water heads of 137 m.

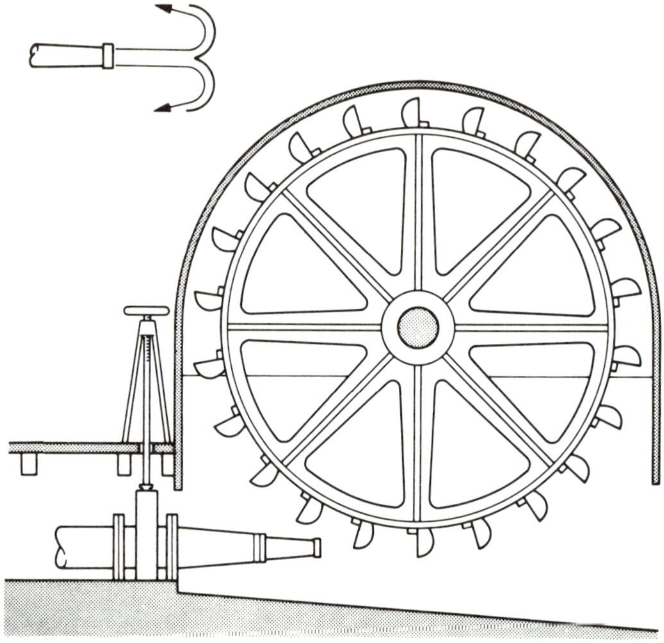

Fig. 57 Pelton wheel

Wind

The windmill as a source of power first appeared in Persia about 700 A.D., but it was not used in Europe before the twelfth century. The early Persian windmill was mounted on a vertical shaft and revolved in a horizontal plane, so it would appear likely that it derived from the Norse water-mill. A similar presumption could be made that the European windmill, with its sails rotating in a vertical plane, was derived from the Vitruvian water-mill.

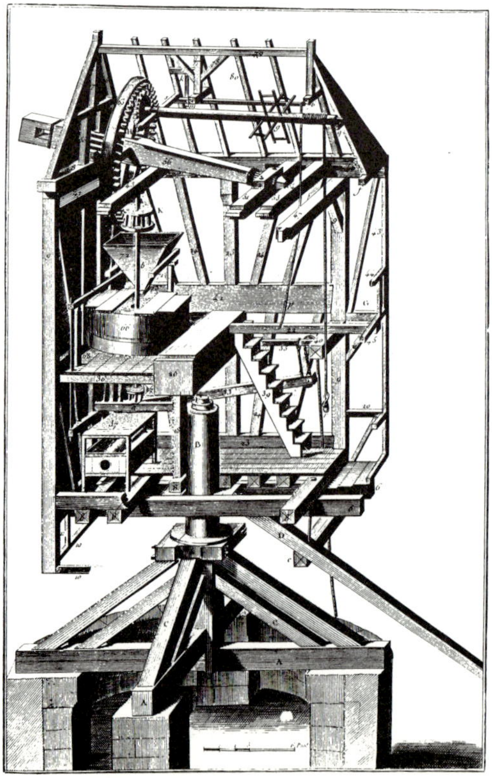

Fig. 58 Fourteen century post mill

Fig. 59 Sixteenth century tower mill

The earliest type of European windmills were the post-mills (fig. 58) so called from the heavy post on which the whole unit was mounted to enable the assembly to be turned into the wind. By the fourteenth century the tower-mill (fig. 59) had been developed; in this mill the building and machinery were stationary, and only the top rotated to carry the sails into the wind. The early sails were of lattice construction with an interlacing of canvas which could be furled to adjust the speed of rotation. Later the canvas sails were replaced by shutters. The heavy and ponderous mechanism was constructed of wood, with the exception of the main bearings for which brass or sometimes stone was used. Heavy grease was employed on all moving parts to reduce the friction.

The sails were turned into the wind manually, an arduous task eased to some extent by the introduction of winders and gearing. The sails were set at a slight angle to the vertical plane to allow them to clear the base of the mill and also to throw some of the weight onto the rear bearing.

In 1745 Edmund Lee invented the fantail (fig. 60) a device which ensured automatically that the sails were kept facing the wind. The fantail consisted of a set of vanes positioned at right angles to the sails. When the sails faced directly into the wind no force was exerted on the fantail vanes, but if the wind altered the fantail was set in motion and through a system of gears brought the sails back into the wind. This was probably the earliest example of the use of automatic control in machinery.

Early windmills were used for grinding corn, but were later used to drive other machinery, including saw-mills. They were used extensively to operate pumps, particularly in Holland where great numbers were used to assist in the reclamation of land from the sea.

Fig. 60 Post mill with fantail

Modifications and improvements in design continued throughout the eighteenth century, and Smeaton included the windmill in his scientific studies, using models to test their efficiency and output. A large Dutch windmill of this period with a sail span of 30 m gave about 7.5 kw in a 40 km/h wind.

The windmill was progressively replaced by the steam engine so that by the middle of the nineteenth century wind could no longer be considered as a major source of power.

Steam

The scientific research which was to lead to the use of steam as a source of power was concerned not with steam but with air. The possibility of the existence of a vacuum, i.e. space without matter, had always been denied by the classical philosophers who relied on the explanation that nature abhorred a vacuum.

TORRICELLI

In 1641, engineers working for the Duke of Tuscany attempted to build a suction pump that would lift water from a depth of 15 m. They found that, despite all efforts, the water would only follow the plunger up the pump to a height of about 8 m. The engineers took this problem to the famous scientist, Galileo, then in the last year of his life. However, he could only make the unhelpful observation that apparently nature's abhorrence ended after the water had been lifted 8 m. The problem continued to occupy him until his death in 1642 when his assistant, Torricelli, continued the research.

Torricelli made the correct assumption that the atmosphere had weight, and that a column of water about 10 m was the maximum that could be balanced by this weight. The fact that the engineers only achieved a height of about 8 m was probably due to the inefficiency of the pump. He also assumed that the weight of the air would only support a proportionally shorter column of a heavier liquid. This Torricelli proved by showing that when a column of mercury in a closed tube was inverted into a bowl of mercury, the height of the column supported by the atmosphere was approximately 71 cm.

The Frenchman, Pascal, carried the experiment further by reasoning that the pressure of the atmosphere decreased the higher one ascended. He proved his theory by having a mercury column carried up a mountain to a height of 1200 m when it was noted that the height of the column had decreased by some 7.5 cm. It was established that the pressure of the atmosphere was approximately 1 kg/cm^2 at sea level, and varied inversely with the height above sea level.

OTTO VON GUERICKE (1602-86)

The effect of atmospheric pressure occupied the attention of numerous reseachers after the work of Torricelli and Pascal had been published. During the 1650s Otto von Guericke, the burgomaster of Magdeburg, carried out his famous experiments, which strikingly demonstrated the effect of atmospheric pressure. He constructed a pump capable of extracting air from a closed vessel and showed how a copper sphere would collapse when the air inside it had been expelled. Probably his most famous experiment (fig. 61) took place

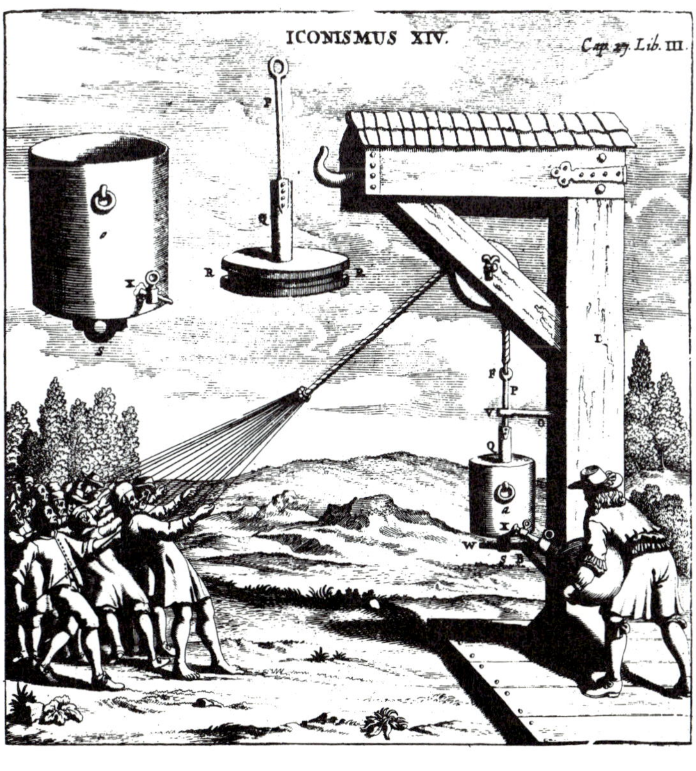

Fig. 61 Von Guericke's demonstration of atmospheric pressure 1654

in 1654. He constructed a cylinder with an accurately fitted piston which was connected to a rope passing over a pulley. The rope was grasped by some fifty strong men, but they were unable to pull the piston more than half-way up the cylinder. Von Guericke then completely evacuated the closed end of the cylinder and the men were unable to prevent the piston descending.

We can see that here von Guericke had the framework of an engine if the vacuum under the piston could be created at regular intervals.

DENIS PAPIN (1647-1714)

Attention now turns to the work of Denis Papin, a French Huguenot. He worked for many years on the problem of creating a vacuum using air pumps and gunpowder. In 1675 Papin came to work with Robert Boyle of Boyle's Law fame. The latter had demonstrated the expansive nature of gases and had suggested that *steam* would quickly expand to fill a space from which the air had been removed. Using this knowledge Papin was able to build a model which consisted of a cylinder and a piston. Water in the cylinder was boiled and the steam produced drove the piston to the top of its stroke. The piston was held in this position while the cylinder cooled and the steam condensed to create a vacuum. When the piston was released atmospheric pressure forced it down the cylinder and weights attached to the piston were raised. Unfortunately Papin appears to have had neither the interest nor the mechanical ability to develop his model into a practical machine.

MINE DRAINAGE

The impetus to harness some alternative source of power to wind and water during the latter half of the seventeenth century derived from the necessity for finding some new method of mine drainage. With the increasing demand for mineral ores, particularly coal, surface deposits had been worked out and it had become necessary to exploit deposits available beneath the earth's surface. As the mines were driven deeper, flooding became an increasing problem which could not be coped with by any available method.

THE STEAM PUMP

The first serious attempt to tackle the problem of drainage using steam power was made by Thomas Savery, a Devonian and erstwhile military engineer with some influence at Court. In 1698 he was granted a patent for 'A New Invention for Raising Water and Occasioning all Sorts of Mill Work by the Impellent Force of Fire'—a grandiose and rather optimistic title. Savery constructed a model of his invention and demonstrated it to the Royal Society in 1699. The essentials of Savery's engine can be readily understood by reference to the simplified diagram (fig. 62). Savery's model had, in fact, two vessels (fig. 63) to which steam was admitted alternately, causing an almost continual flow of water up the force pipe.

Although theoretically sound, in practice Savery's atmospheric pump (for as it had no moving parts it could not be classed as an engine) was unreliable, inefficient, and dangerous. He found that he could only lift water from a depth of about 6 m. Thus with the height of the force pipe limited by the steam pressure available from his primitive boiler to about 7.5 m, the total

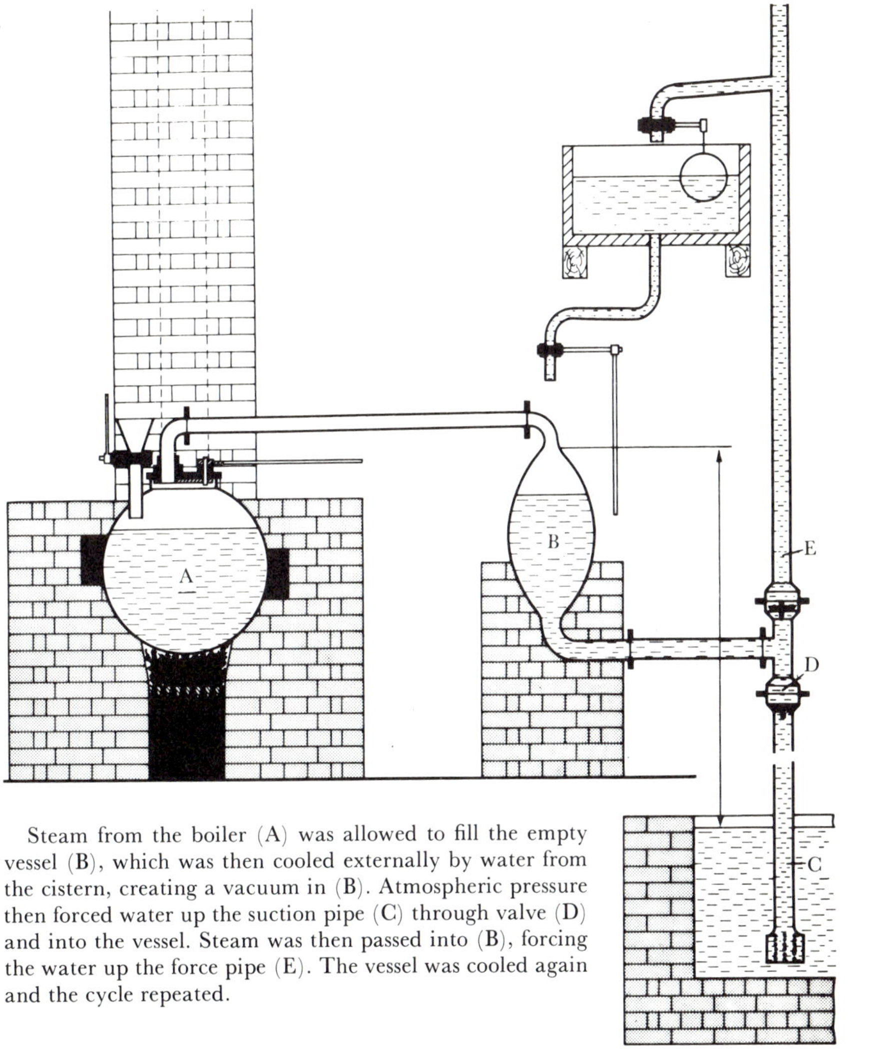

Steam from the boiler (A) was allowed to fill the empty vessel (B), which was then cooled externally by water from the cistern, creating a vacuum in (B). Atmospheric pressure then forced water up the suction pipe (C) through valve (D) and into the vessel. Steam was then passed into (B), forcing the water up the force pipe (E). The vessel was cooled again and the cycle repeated.

Fig. 62 Principle of the Savery steam pump 1698

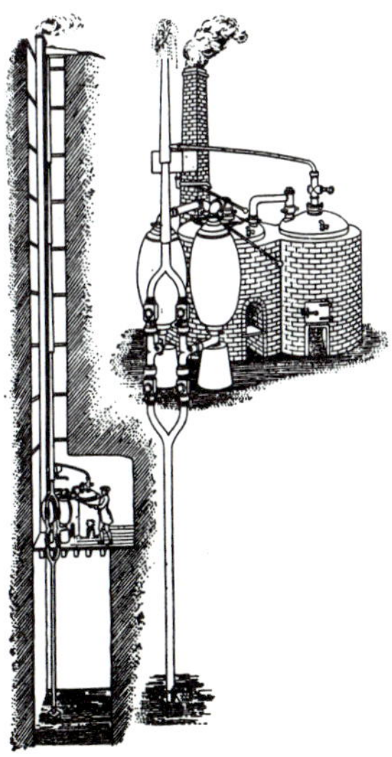

Fig. 63 Savery's engine

lift of about 15 m made the apparatus of little use in solving the problem of mine drainage. It is doubtful whether one of Savery's 'engines' was ever installed in a mine although some were installed in large houses to supply domestic water. Given the technical knowledge and skill of the nineteenth century, Savery's engine would have fulfilled his expectations, but the solution to the problem at the time lay in a different direction.

There exists some doubt regarding the originality of Savery's device, certainly other inventors experimented with the use of steam during the seventeenth century. Edward Somerset, the Marquis of Worcester (1601-67) has been credited with the invention of an apparatus similar to Savery's, but the documentation is too vague to afford him more than a passing mention in this book.

THE ATMOSPHERIC ENGINE

The first engine to be used successfully for the operation of mine drainage pumps was designed and built by Newcomen, a blacksmith and ironmonger of Dartmouth in Devon. By the nature of his trade he had frequent contact with the tin mines, and was well aware of the inadequacy of the horse-powered pumps that were then used to drain the mines. His work appears to be a development of the ideas of Denis Papin, but there is no evidence to show that he, in fact, had knowledge of Papin's work.

Newcomen, together with his assistant, John Calley, who is described as a plumber, laboured for ten years to perfect his engine. Probably the most significant incident on their road to success was the discovery of the method of injecting water into the cylinder in the form of a fine spray to ensure the rapid condensation of the steam under the piston. Newcomen's engine is illustrated in the simplified diagram (fig. 64).

Steam from the hemispherical boiler, which was derived from the brewer's kettle, passed into the cylinder through the steam valve to fill the cavity beneath the piston. The steam valve was then closed and a spray of water was admitted to the cylinder by opening the injection cock. The steam was rapidly condensed by the cold water, leaving a vacuum beneath the piston. Atmospheric pressure, acting on the open side of the piston, then forced the piston down the cylinder to produce the power stroke.

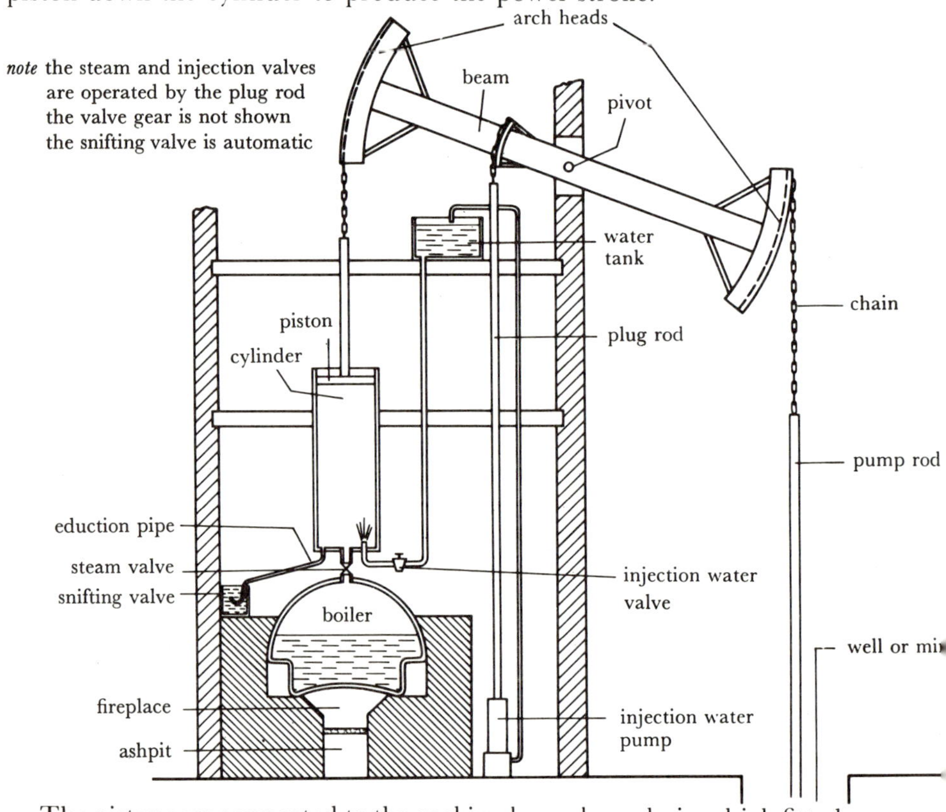

Fig. 64 The Newcomen engine

The piston was connected to the rocking beam by a chain which fitted over an arch head designed to restrict the movement of the piston rod to a vertical plane. As the beam was pulled down the pump rod was lifted and water raised.

After the completion of the downward stroke of the piston, steam was re-admitted and the piston rose, mainly owing to the weight of the pump rod and chain acting on the opposite end of the beam. The expanding steam forced the condensed water out of the cylinder via the eduction pipe, and at the same time expelled through the snifting valve any air which had accumulated. The snifting valve, so called because of the noise it produced, was simply a tube which led from the base of the cylinder to a cistern of water. Air forced from the cylinder bubbled up through the water in which the tube was immersed to form a simple non-return valve. This device prevented the engine from becoming 'wind logged', the name given to a phenomenon which had caused Newcomen much trouble in his first model and had tended to stop the engine after a dozen or so strokes. Newcomen's ingenious solution to the problem was a measure of his great mechanical ability.

It will be remembered that no machine capable of boring an accurate cylinder existed in 1712, and to produce an airtight seal around the piston Newcomen devised the method of having a layer of water above the piston. The first Newcomen engine was erected in the vicinity of Dudley Castle, near Wolverhampton, in 1712 (fig. 65). The engraving of this engine which was made in 1719 shows that the injection cock and steam valve were operated automatically by a rod which was connected to the beam.

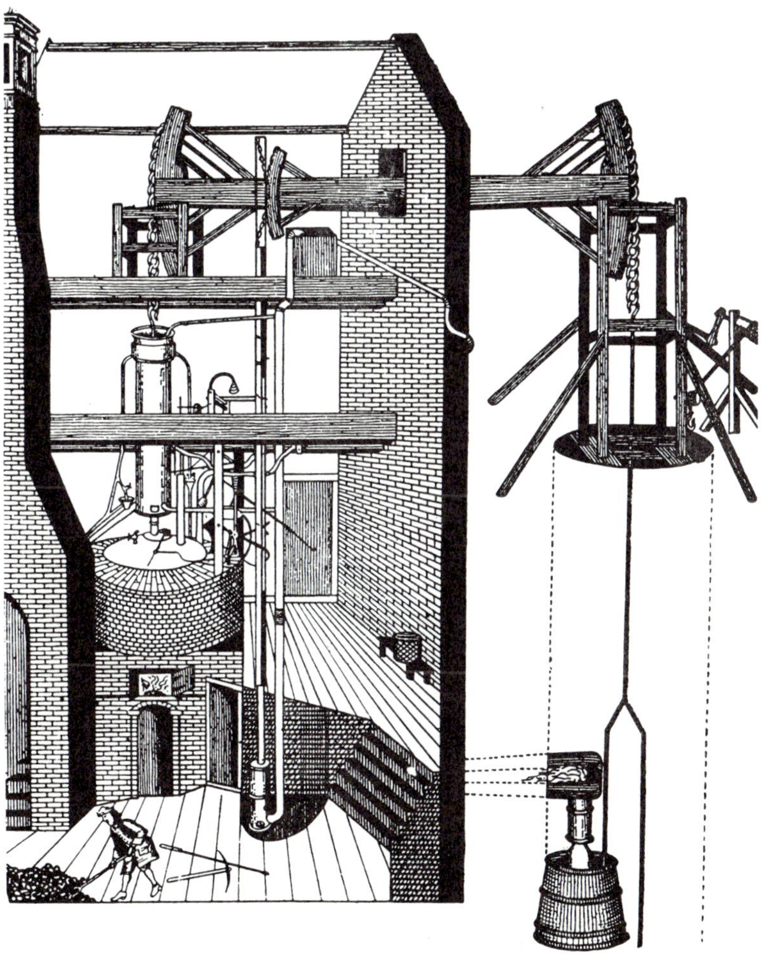

Fig. 65 The first Newcomen engine built 1712

As the patent taken out by Savery in 1698 was held to cover all steam engines, Newcomen was forced to take him into partnership before his engine could be marketed. Thus Savery, who made no contribution to the development of the engine, has acquired some undeserved credit over the course of the years.

Although the prestige of James Watt has tended to obscure the work of Newcomen, modern historians now give him full credit for the invention of the steam engine, or, more correctly, the atmospheric engine. The fact that some Newcomen engines were in constant use for upwards of a hundred years is sufficient testimony to the genius of this great inventor.

Newcomen lived to see his engine in use in places as far afield as Sweden, Russia, and Hungary. After his death in 1729, and with the expiration of the patent in 1733, the way was open for anyone to construct the engine. Cylinder sizes increased to as much as 1.8 m diameter in some cases, and a way was found of reducing costs by using cast iron rather than brass cylinders.

The availability of coke-produced cast iron, and improved methods of casting due to Abraham Darby were discussed in chapters 1 and 3. The cost of the engine was still high—a figure of £1700 is quoted for an installation in 1745. Moreover, with a coal consumption of up to three tonnes a day, such engines were an uneconomic proposition for duties other than coal mine drainage. In consequence the engine proved especially popular in the coal mining areas of the North of England where it could be run on stocks of unsaleable coal. In the tin mining districts of Cornwall it was a much less attractive proposition, for the coal used had to be transported from the distant coalfields, and was subjected to a tax until 1741.

JOHN SMEATON (1724-92)

The men who built the engines during the thirty years after the death of Newcomen were mainly practical engineers with little understanding of the theoretical concepts of his design. To them, increased power appeared synonymous with larger cylinders, and they paid little regard to mechanical or thermal efficiency. John Smeaton, a Yorkshire engineer, undertook a serious study of the Newcomen engine, and in 1765 constructed an experimental model to help him in his work. He also undertook a survey of a number of existing engines, and from the data collected he devised a method of comparing the work done by these engines. This he called the 'duty', which was the amount of water in millions of pounds which could be raised one foot (30.5 cm) by the burning of one bushel, or 74 lb (33.6 kg), of coal. He found that the engines with the largest cylinders were not necessarily the most efficient. Poor performance, he claimed, could be due to serious defects in engine construction. Inaccurately bored and badly proportioned cylinders, ill-fitting pistons, poorly timed valve gears, and inadequate boilers could all be contributing factors.

Taking account of his observations, Smeaton constructed an engine for Long Benton colliery, Northumberland, in 1772 which included many refinements. This engine had a duty of 9.45 millions compared with a previous best performance of 7.44 millions (i.e. using one bushel of coal it would raise 9,450,000 lb of water through one foot). To improve the quality of cylinders Smeaton designed a boring machine for use in the Carron iron foundry, which produced a *true* circular bore. Although his boring mill was a big advance on previous methods it did not guarantee a parallel bore; this only became possible with the later invention of John Wilkinson which is discussed in chapter 4. Naturally much thought had been given to extending the scope of the Newcomen engine by using it to produce rotary motion. The obvious solution was to use a connecting rod and crank, a device which was well known, but Smeaton, in common with other engineers of the time, reasoned incorrectly that the variable stroke of the engine would smash the connecting rod. Smeaton's answer to the problem was to use the Newcomen engine to pump water from the bottom to the top of the mill race, using the water-wheel to provide rotative power. It was exactly this technique which was used by Abraham Darby II at Coalbrookdale.

JAMES WATT (1736-1819)

Some thirty-five years elapsed after the death of Newcomen before the first fundamental change in the design of the steam engine occurred. James Watt, working as a mathematical instrument maker at Glasgow University, was asked to repair a model of a Newcomen engine belonging to the Department of Natural Philosophy. He decided that the difficulty he experienced in making the engine run for more than a few strokes was due to the inability of the boiler to provide the amount of steam required for continuous working. More important was his realisation that the high consumption of steam was due to the cooling of the cylinder by the injection of water after every stroke. After studying the problem for some months the idea came to him 'that as steam was an elastic body it would rush into a vacuum, and if a communication was made between the cylinder and an exhausted vessel, it [the steam] would rush into it and might be there condensed, without cooling the cylinder'.

By constructing a model, Watt was able to test the practicability of his idea. Originally he had intended to incorporate the condenser, the name he gave to the 'exhausted vessel', in an inverted direct-acting engine. His limited knowledge of engineering, however, led him to change plans and apply his invention to the existing Newcomen engine.

Figure 66 shows the engine as it first appeared in its commercial form, manufactured by the firm of Boulton and Watt of Soho, Birmingham. In order to keep the cylinder of the engine as hot as possible he enclosed it in a jacket so that steam could circulate between the jacket and the cylinder. As the piston rod now had to pass through the steam jacket he invented the stuffing box, a device to prevent steam escaping past the piston rod. To remove the condensed steam and air from the condenser the latter was connected to an air pump operated by the beam. Both the air pump and condenser were situated in a tank of cold water and the condensate passed to a hot well which provided preheated water to the boiler. The cycle of operations can be followed with reference to the simplified line diagram (fig. 67).

Thanks to Wilkinson's timely invention of the boring machine the first two Boulton and Watt engines were installed in 1776, one of them being for the famous ironmaster himself. Boulton and Watt sold their engines on a premium basis, charging one-third of the value of the fuel saved. The firm supplied important items such as the valves, and the cylinder was supplied by Wilkinson. The remaining parts of the engine were manufactured locally, and erection took place on site, supervised by an employee of Boulton and Watt.

By 1780 Watt was working on the conversion of his engine to provide rotary power, spurred on by Boulton, who was fully aware of the demand. To pro-

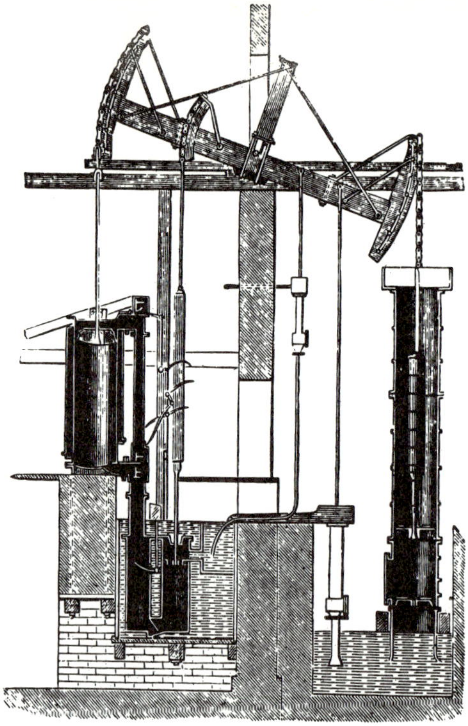

Fig. 66 The Watt single acting engine

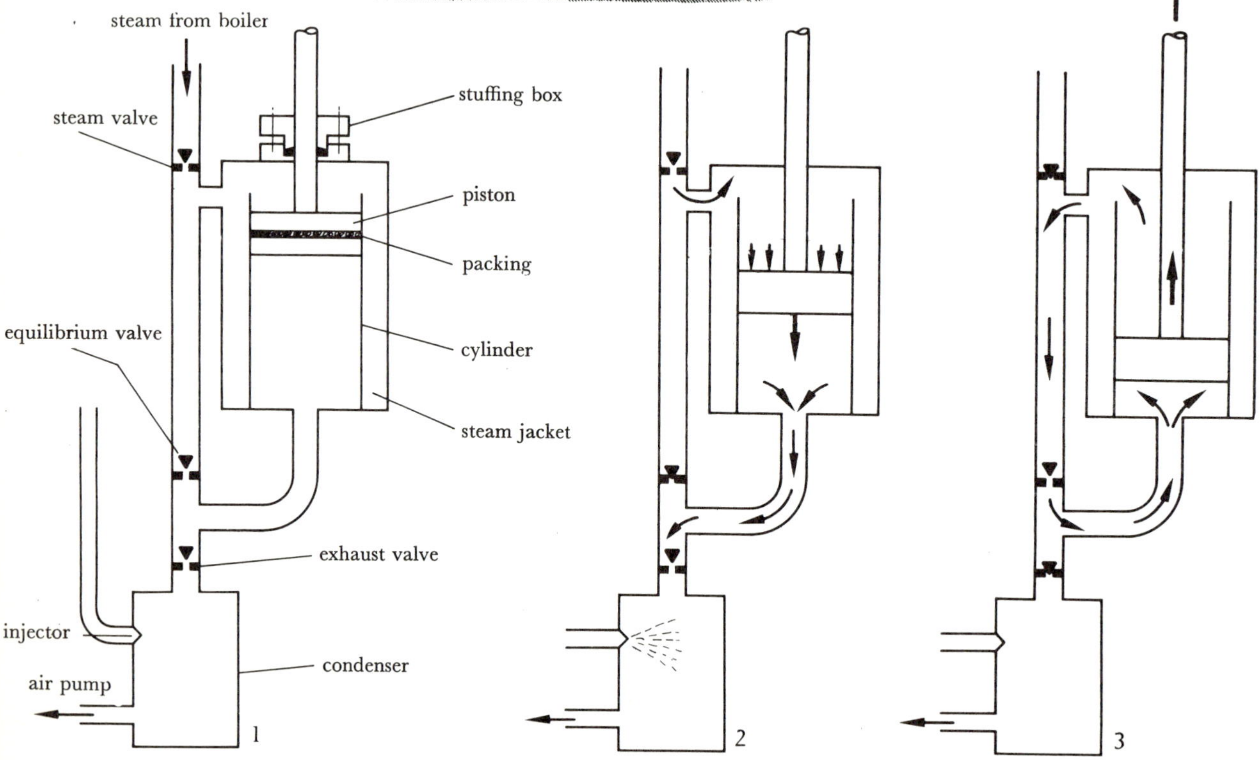

Fig. 67 How the early Watt engine operated

Beginning with the piston at the top of its stroke (1), all valves were opened to allow the steam to circulate and expel any air. The equilibrium valve was closed (2) and water injected into the condenser condensed the steam which passed from the cylinder through the exhaust valve, creating a vacuum below the piston. Steam at little over atmospheric pressure acting on top of the piston produced the downward power stroke. On completion of the power stroke the steam valve and exhaust valve were closed and the equilibrium valve was opened. Steam passed from the top to the bottom of the cylinder via the equilibrium valve, equalising pressure and allowing the piston to be raised by the weight of the pump rods on the other end of the beam.

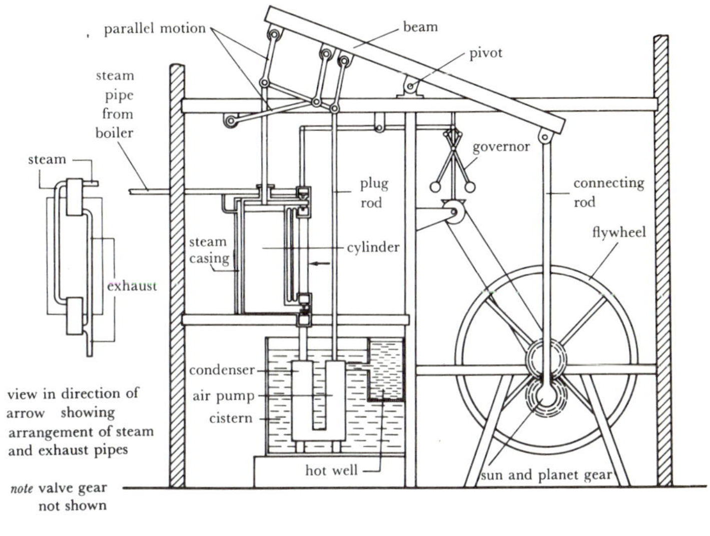

Fig. 68 The Watt double acting engine

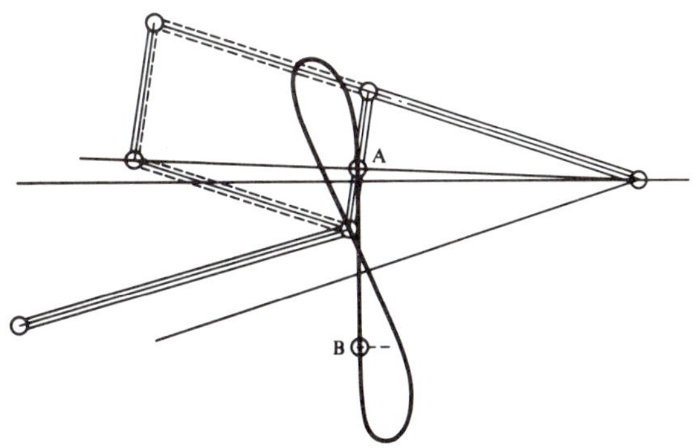

Fig. 69 Diagram of Watt's parallel motion 1784

duce rotary motion it was desirable that the engine should be double acting, in other words that both up and down strokes should be power strokes. Although this in itself was not too difficult to achieve, it required that the piston rod be rigidly connected to the beam. As the end of the beam described an arc when in motion, Watt designed his famous linkage to keep the piston rod travelling in a true vertical path. The linkage can be seen in the print of the double acting rotary engine (fig. 68) and the line diagram (fig. 69) shows the locus of the end of the piston rod. Unfortunately, James Pickard of Birmingham had secured a patent in 1780 which covered the use of the crank and connecting rod, and so Watt was forced to design an alternative method for converting reciprocal motion to rotary motion. This he did with his sun and planet mechanism which comprised a planet-gear fixed to the lower end of the connecting rod which meshed with and imparted rotary motion to another gear (the sun). To carry the gearing over the top and bottom dead centres, a large flywheel was incorporated. The first double acting rotary engine was built in 1783, again for John Wilkinson, and it continued to be manufactured in this form until the expiration of Watt's patent in 1800.

James Watt continued to work on the theory of steam engines. He discovered the method of cutting off the steam supply early in the cycle to make use of the expansive nature of steam, thus economising on fuel. To measure the work done by the engine he invented the indicator which, in a more refined form developed by John Southern, produced a closed diagram to scale on a sheet of paper. By measuring the area of the diagram using a planimeter the work done could be calculated. Watt gave a value of 33,000 foot pounds per minute to a unit which he called one horse power (h.p.) to describe the work done by an engine.

Watt was highly suspicious, allowing no one to construct the engine under licence or to modify it in any way. The sheer volume of orders, however, obliged his firm to use other engineers as their agents, invariably the most able mechanics in the country. In Lancashire, the Manchester firm of Sherratt and Bateman probably constructed more engines than Watt and Boulton themselves. Like the other firms used in this way they assisted in disseminating Watt's exacting standards of precision over a wide area.

High-Pressure Steam

It was known that steam raised to well above atmospheric pressure could be used to drive down the piston of the steam engine directly. This was an alternative to using steam as an agency for the creation of a vacuum and it would substantially increase the power of the engine. As early as the 1720s, Jacob Leupold, a German, had produced a design for a high-pressure steam engine (fig. 70). The major problem to be surmounted was the construction of a boiler capable of withstanding the increased pressure, a rock upon which Savery had foundered. It was only during the later years of the eighteenth century, when wrought iron plate in large enough sizes became available, that the construction of suitable boilers became a practical proposition. William Murdock, an employee of Boulton and Watt, designed and made a model of a high-pressure engine in 1785, but Watt, who considered steam at much above atmospheric pressure to be highly dangerous, discouraged further development.

RICHARD TREVITHICK (1771-1833)

Trevithick, a Cornish mining engineer, began experimenting with high-pressure steam in 1798. His first problem was to design a suitable boiler, and by 1802 he had arrived at the form shown in fig. 71. The cylindrical portion was cast in iron with an end plate of wrought iron to which was bolted the firebox and flue. The end plate could be removed to allow the boiler to be cleaned and the vertical cylinder was sunk into the boiler to keep it hot. The steam was admitted, as in the Leupold model, through a four-way cock to alternate sides of the piston to make the engine 'double acting', and was fully exhausted directly to the atmosphere. The engine was direct acting, that is, the piston rod was connected by means of a crosshead onto the connecting rod, dispensing with the beam. This reduced the size of the engine considerably, and encouraged Trevithick in his ambition to utilise his engine to drive a moving vehicle. By 1802 he had run a steam carriage successfully, and in 1804 came his historic locomotive—the first in the world. It was displayed at Pen-y-daren, South Wales, where the locomotive pulled wagons carrying ten

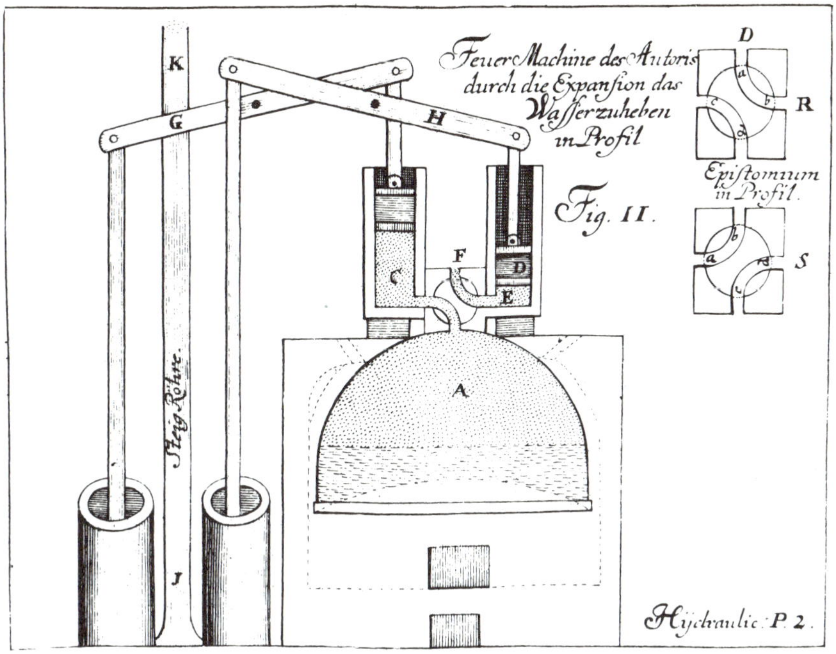

Fig. 70 Leupold's high pressure steam engine

Fig. 71 Trevithick's high pressure steam engine 1802

tonnes of ore and seventy men for 14.5 km at a speed of 8 km/h. His ambitions in this direction, however, were frustrated mainly because of the inadequate roads and inferior rail tracks. More immediately significant were Trevithick's stationary engines which were built under licence throughout the country. They were used to power a variety of industrial processes, including sugar-milling, iron-rolling, and corn-grinding.

OLIVER EVANS (1755-1819)

In America a self-educated farm boy, Oliver Evans, was also experimenting with high-pressure steam. In 1802, when mechanical knowledge in the U.S.A. lagged behind that of Britain, he constructed a successful direct double-acting steam engine using steam at a pressure of 3.5 kg/cm^2. The steam was applied from a boiler designed by Evans which anticipated the fire-tube boiler of a much later date. The cylinder was a mere 15 cm in diameter with a piston stroke of 45.72 cm. Evans solved the straight-line motion problem by using a 'walking beam'. From the peculiar loping motion of the beam stemmed the name 'grasshopper' engine.

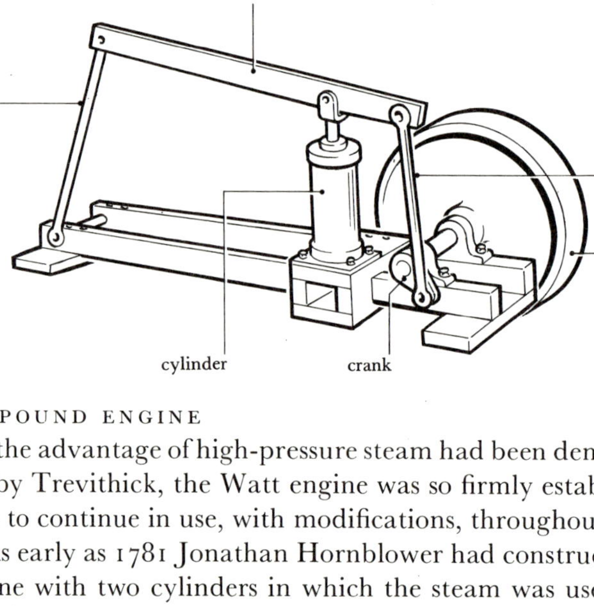

Fig. 72. Evans' high-pressure steam engine

THE COMPOUND ENGINE

Although the advantage of high-pressure steam had been demonstrated convincingly by Trevithick, the Watt engine was so firmly established by 1800 that it was to continue in use, with modifications, throughout the rest of the century. As early as 1781 Jonathan Hornblower had constructed a modified Watt engine with two cylinders in which the steam was used successively. The steam first passed to a high-pressure and then to a low-pressure cylinder. This later became known as 'compounding' or 'two-stage expansion', but Watt succeeded in suppressing Hornblower's engine after a prolonged legal struggle.

After the expiry of Watt's patent, Arthur Woolf was granted a patent for a compound engine in 1803; despite early setbacks he constructed a successful engine with a ratio between the H.P. and L.P. cylinders of 1 : 5. The two cylinders stood side by side and the pistons were connected to work in tandem at one end of the beam. Trials showed a saving in fuel of up to 50 per cent over the Watt engine, but surprisingly the engine had little success in this country although it became increasingly popular on the continent, probably owing to the higher cost of fuel there.

As the century progressed, factories and mills were demanding more power than could be obtained from the beam engines installed. John Mcnaught, a Lancastrian, introduced in 1845 the method of compounding existing engines by adding a high-pressure cylinder which operated on the opposite end of the beam to the low-pressure cylinder. This practice, known as 'Mcnaughting', established the use of the compound engine in Britain as an economic alternative to the expensive solution of completely replacing old beam engines.

Compounding enormously improved thermal efficiency. Where the Newcomen engine used 30 lb of coal for every horsepower hour, Watt used $7\frac{1}{2}$ lb and the average mid-nineteenth-century compound engine used $2\frac{1}{2}$ lb. With the promise of fuel saving on this scale, steam power appeared increasingly more advantageous to the entrepreneur.

THE CORNISH ENGINE

Concern over the declining efficiency of the Watt engine installed in the tin mines of Cornwall led to a systematic study of all the engines in order to effect an improvement. Trevithick built a modified Watt-type engine which retained the condenser but made use of high-pressure steam. This engine, installed in 1812, was the prototype of what came to be known as the Cornish engine. The efficiency of these engines was progressively improved until, by 1842, Taylor's Cornish engine achieved a duty of 107 millions. Cornish engines remained in use throughout the nineteenth century, particularly in mines and waterworks where they acquired a reputation for reliability.

OTHER DEVELOPMENTS IN THE EARLY NINTEENTH CENTURY

The firms of Matthew Murray in Leeds and Henry Maudslay in London were serious rivals to Boulton and Watt Ltd in the early years of the nineteenth century. Much of their work on machine tools discussed in chapter 4 was a by-product of their work in the production of steam engines.

Murray made some notable contributions to the development of the steam engine. He introduced the practice of mounting the complete engine on a cast iron bed-plate, and assembling the engine in the factory. He perfected the D-type slide valve introduced by Murdoch in 1799. This simplified the valve gear, and in conjunction with the eccentric, a mechanism designed to give motion to the slide valve, formed the basis of all future valve gear design. Murray developed an inverted beam engine known as the side-lever engine which remained in use as a marine engine until the middle of the century.

Maudslay introduced his table engine in 1807. This compact engine with its vertical cylinder mounted on four cast iron legs achieved great popularity. Maudslay dispensed with the beam and used a crosshead to drive the crankshaft via a connecting rod. It says a good deal for the reputation of Maudslay Sons and Field that their marine engines were chosen for the ships of I. K. Brunel, one of the most exacting engineers of the nineteenth century.

BOILERS

To meet the demands of the high-pressure steam engine, it was essential that

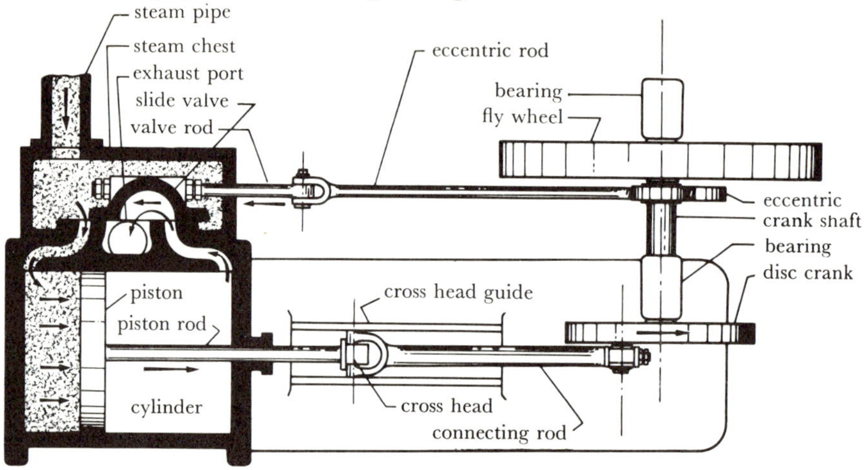

Fig. 73 Steam engine showing slide valve, eccentric and crosshead

stronger and more efficient boilers were developed. We have already seen the transition from the simple copper kettle-type boiler of the early Newcomen engine to the cylindrical cast iron variety constructed with internal firebox, developed by Trevithick. The most important step was taken by William Fairbairn whose Lancashire boiler contained two flues to increase the heating surface. Further development consisted of a number of tubes placed through the boiler shell to increase the heating area. These boilers fell into two main types—water-tube and fire-tube. In water-tube boilers the water passes through the tubes which are surrounded by the heat from the fire, while in fire-tube boilers heat from the fire passes along the tubes which are surrounded by water. A fire-tube boiler was used by Robert Stephenson in the *Rocket*, and this type has been used universally for steam locomotives ever since.

The development of the steam engine continued rapidly throughout the nineteenth century, so that by 1900 the successors to the Newcomen engine were triple expansion condensing engines with forced lubrication, developing 2000 h.p. By this time, however, the steam engine was due to be eclipsed by another source of power derived from steam—the steam turbine.

STEAM TURBINES

The desirability of using steam to produce direct rotary motion was appreciated quite early in the history of steam engines. In the reciprocating engine the conversion of reciprocative motion into rotary motion using the crosshead, connecting rod, and crank, consumed a considerable part of the energy of the steam. The solution lay in the development of the steam turbine which had been demonstrated by Hero of Alexandria in the first century A.D. Hero's aeolipyle was, in fact, a reaction turbine. Steam was supplied from a boiler to a hollow sphere mounted on trunnions. The steam from the sphere was discharged through two diametrically opposed bent pipes and caused the sphere to revolve. Hero regarded his aeolipyle solely as a toy, but its potentialities as the basis of a working machine were obvious.

In 1629 Giovanni Branca made a drawing (fig. 74) of an idea for using a steam turbine to do useful work. It was, in fact, impracticable considering the mechanical limitations of the time, but it serves to illustrate an alternative form of turbine—the impulse turbine. All turbines are either of the impulse or reaction type, or a combination of both. Motion to the impulse turbine is imparted by the changing momentum of the steam as it strikes the curved blades, whereas the reaction turbine is governed solely by Newton's Third Law which states 'that to every action there is an equal and opposite reaction'.

Richard Trevithick built a full-size reaction turbine which had steam nozzles mounted on the end of a rotary pipe 4.5 m long. He referred to it as a 'Whirling Engine', but he was defeated by his inability to get the steam to leave the nozzles at a sufficiently high velocity.

Fig. 74 Branca's steam turbine

CHARLES ALGERNON PARSONS (1856-1931)

The development of the first practical steam turbine was the work of the Hon. C. A. Parsons who was inspired by the problem of driving the newly-introduced electric generator at a speed beyond the capabilities of the reciprocating steam engine.

Parsons's turbine was of the axial-flow reaction type. It consisted of fourteen circular rows of blades mounted on a rotor which revolved inside a casing or cylinder (fig. 75). Fixed to the inside of the cylinder were blades which projected inwards, mounted in circular rows which fitted between the blades on the rotor; a small clearance was provided between the fixed and moving blades. The cylinder increased in diameter from the high pressure end so that the expansion of the steam took place gradually, making possible slower speeds of rotation than those of impulse turbines. In his original steam turbine Parsons made the rotor in two parts, and arranged for the steam to enter at the mid-point of the cylinder and flow outwards, thus eliminating end-thrust. To prevent vibration at the high speeds required, he designed special floating end-bearings to enable the shaft to find its own centre of rotation. The turbine drove a two-pole dynamo which produced 7.5 kilowatts at 100 volts for a turbine speed of 18,000 r.p.m.

Fig. 75 Parson's turbine

Parsons first showed his machine at an exhibition in 1885, but attracted little serious attention. The Parsons turbine was first installed in a public power station in Newcastle in 1888, it supplied 75 kw and was built by Clark, Chapman and Co., the firm in which Parsons was a partner.

Following a dispute with his partners, Parsons had to relinquish control of his patent on the axial-flow turbine in 1889 and was forced to continue his experiments using the radial-flow type of turbine. With this turbine he introduced the use of the condenser which reduced the steam consumption considerably so that the turbine used less steam than a reciprocating engine of comparable output.

In 1893 Parsons regained control of his original patent and very soon produced improved axial-flow turbines with outputs rising to 1000 k.w. He then turned his attention to ship propulsion using steam turbines, and in 1897

Fig. 76 The Turbinia

the Turbinia, powered by three turbines developing 1500 kw appeared unannounced at the Jubilee Naval Review. This small vessel, only 30 m long and of 44 tonnes displacement, astonished the onlookers by steaming through the assembled ships at 35 knots, easily outrunning the destroyers sent to intercept her. So impressed were the Admiralty that they ordered two new destroyers to be equipped with Parsons's turbines, and from then on the steam turbine gradually replaced the reciprocating engine in ships.

GUSTAV DE LAVAL (1845-1913)

At the same time as Parsons was developing his axial-flow turbine, de Laval, in Sweden, was working on a steam turbine of the impulse type roughly analogous to the water-driven Pelton wheel mentioned earlier.

Steam directed on to the curved blade through four or more nozzles (fig. 77) caused the motor to spin. The section through one of the nozzles shows how the opening in the nozzle expanded towards the mouth. This feature allowed the steam to expand and increased its velocity so that it struck the blades at a speed of 900 to 1200 m/s. Impulse turbines of this type ran at speeds varying between 30,000 r.p.m. for small turbines down to 9000 r.p.m. in the largest models.

At these high speeds balance and vibration were important problems and to overcome them de Laval used a long slender shaft to carry the rotor. One end of the shaft was mounted in a plain bearing while the other was mounted in a spherical bearing. With this arrangement the shaft settled down to vibrationless running at the design speed of the turbine after passing through short periods of vibration at critical speeds in the lower ranges.

Curtis, an American, improved on the de Laval turbine by making use of the remaining energy of the steam in a further series of rotors. His original machine was vertical and the turbine wheels were separated by rows of guide vanes to change the direction of the steam.

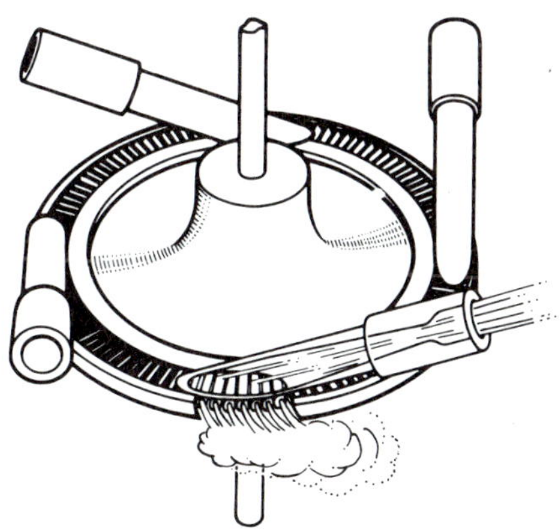

Fig. 77 De Laval's turbine

The Economic and Social Effects of Steam Power

At the height of its power in 1835 one writer could describe steam as 'the controller general and main spring of British industry, which urges it upwards and never suffers it to lag or loiter'. An early example of the utilisation of steam power was in the textile industry, where the pattern had been a rather primitive manufacturing organisation based on the 'domestic system'.

The spinning and weaving was undertaken by agricultural labourers as outwork to supplement their normal income, and this situation was not radically changed by the increasing use of water power.

In 1784 an Oxford don, the Rev. Edmond Cartwright, built a weaving loom which could be driven by a steam engine and a few years later in 1790, Crompton's mule, which was a multispindled spinning machine, was also adapted to make use of this new source of power. This meant that by the 1820s steam engines of 2200 kw were being worked in the cotton industry, and since it was not possible to locate such monsters in the living room of cottages, steam power sounded the knell of the domestic system in favour of the new factories or mills.

In other chapters the adoption of the steam engine in iron and steel rolling mills has been recounted, and it has also been shown that steam was of great importance in mining. 'The greatest service that the heat [steam] engine has rendered in England', wrote one foreign commentator 'is without doubt the fresh stimulus given to the exploitation of her coal mines which were languishing and in danger of collapse as a result of the ever-growing difficulties of controlling and removing water'.

More steam-driven machinery had the obvious effect of increasing the demand for iron and coal and of creating, in consequence, an upward spiral of general productivity. Outside industry the influence of steam was no less marked. In 1814 a German steam-powered cylinder machine was used to print newspapers, and in so doing made a significant contribution to the growing movement for democracy. In another way the introduction of the Perkins steam-tube oven and the steam threshing machine in the middle of the last century helped to provide cheaper food.

But above all, it was in transport that the full splendour of steam power could be shown. The first experiments of Trevithick which have been described in this chapter were followed by the more persistent, if less brilliant

Fig. 78 Hedley's *Wylam Dilly*

pioneer work in the north-east of England. The *Wylam Dilly* of William Hedley and the *Puffing Billy* of Timothy Hackworth encouraged another north-easterner, George Stephenson, to design his first engine, the *Blücher,* in 1814. By 1825 Stephenson had completed the first railway link in the world—from Stockton to Darlington—and had a powerful interest in a company which was to produce many of the famous early engines like the *Locomotion* (1825), the *Rocket* (1829), and the *Planet* (1830).

The development of the steam locomotive is a serious study in itself, and it must suffice here to note briefly some of the main stages. The multitubular boiler, whose working has already been explained, was first adopted for the locomotive by Marc Seguin of France, and later by Robert Stephenson in Britain.

It will be remembered that greater efficiency was obtained when the steam engine was compounded and this technique was in time also applied to the locomotive. In Britain Francis Webb compounded with two external cylinders, and in the 1880s A. G. de Glehn, a French engineer, designed a model working on four cylinders.

Steam also had a significant part to play in water transport. The earliest successful steam boat was French—the *Pyroscaphe*—built in 1783 by the Marquis d'Abbans. In 1788 a boat constructed by William Symington and propelled by an atmospheric steam engine made its trial run on a Scottish lake. A later version of this boat was seen by the American inventor, Robert Fulton, who constructed in 1807 a paddle steamer—the *Clermont*—which may be considered as the first commercial success in this field. But there were formidable difficulties to be met, first in distributing the weight of a large engine in constant motion, and secondly, in transporting the large stocks of fuel which were necessary. These problems encouraged experimentation to find a way of building a compact engine which could be run with the maximum economy.

In 1815 Maudslay built his first marine steam engine which worked on the side lever principle to impart greater stability. Another development was the oscillating engine first used in 1822 which had the advantage of direct working without a connecting rod.

The ancient theory of the Archimedian screw incorporated in the propeller and combined with steam power provided another means for propelling ships, and in 1830 the first steamship of this type was launched. In the late 1850s it was found more beneficial to invert completely the steam engine in ships, and in the 1870s compounding was introduced, which led to a triple-expansion marine engine.

Steam transport accelerated social as well as economic changes in all manner of ways. The railway now transported perishable foods like fish and milk to the large towns. Cheap fares brought travel and holidays within the budgets of ordinary people and prompted the great expansion of the seaside resorts at the end of the nineteenth century.

Steam, in fact, with its influence on so many aspects of life, amply substantiated a remark made by Matthew Boulton to a visitor to the Soho works: 'I sell here, sir, what all the world desires to have—power.'

Additional Reading

H. W. DICKINSON, *A Short History of the Steam Engine,* 2nd edn., Cass 1963.
R. J. LAW, *The Steam Engine,* H.M.S.O. for the Science Museum 1965.
PROJECT TECHNOLOGY HANDBOOK, Industrial Archaelogy of Water-mills and Water-power, Heinemann Educational Books/Schools Council 1975.

Chapter 6
Sources of Power 2

Hydraulics

The French scientist, Pascal, whose work in relation to steam power has already been discussed, carried out in 1647 a series of experiments with fluid machines. From these experiments he made the important deduction that 'in a fluid at rest pressure is exerted equally in all directions', and constructed a machine to illustrate his proposition. This machine showed that when pressure was applied to the liquid by the piston in the small cylinder this pressure acted uniformly on the underside of the large piston. Thus, if the area of the small piston was one square cm and the area of the large piston 100 square cms a force of one newton applied to the small piston would cause a force of 100 newtons weight to act on the large piston. Pascal, however, did not construct a useful machine and it was not until Joseph Bramah invented the hydraulic press in 1796 that his ideas gained practical expression.

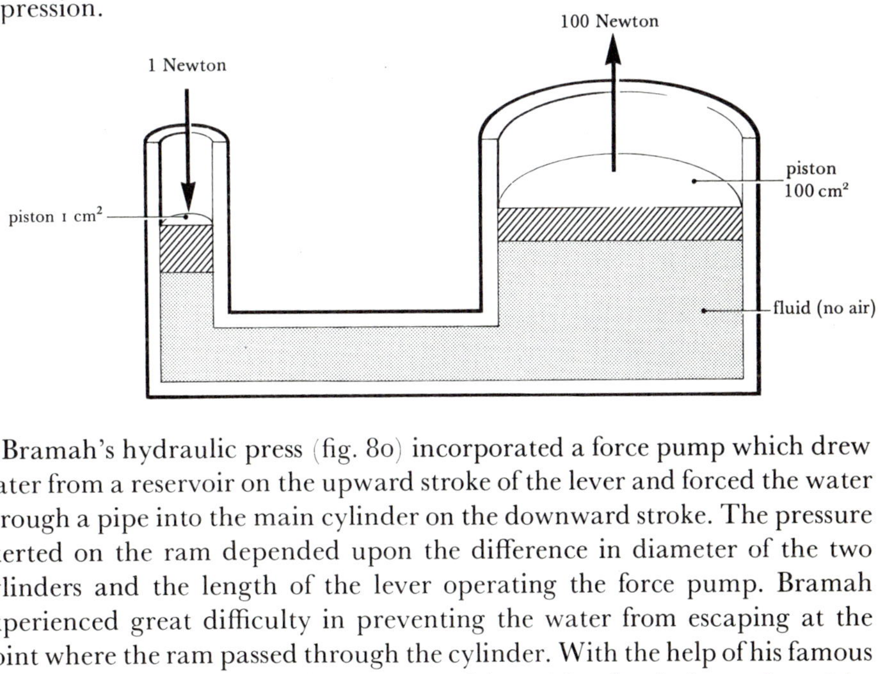

Fig. 79 Principle of the hydraulic press

Bramah's hydraulic press (fig. 80) incorporated a force pump which drew water from a reservoir on the upward stroke of the lever and forced the water through a pipe into the main cylinder on the downward stroke. The pressure exerted on the ram depended upon the difference in diameter of the two cylinders and the length of the lever operating the force pump. Bramah experienced great difficulty in preventing the water from escaping at the point where the ram passed through the cylinder. With the help of his famous assistant, Henry Maudslay, he overcame this problem by the invention of the leather-cup washer. This was inserted between the ram and the cylinder in such a way that the increasing pressure forced the inverted U-shaped washer to press tightly against the ram and so prevent leakage (fig. 81). The first uses of the press were for wool baling and seed crushing, but its potentialities for metal forming were soon appreciated, and hydraulic presses were designed for such heavy jobs as flanging boiler end plates and forging iron sheets.

The press embodied all the requirements of a hydraulic jack and it was

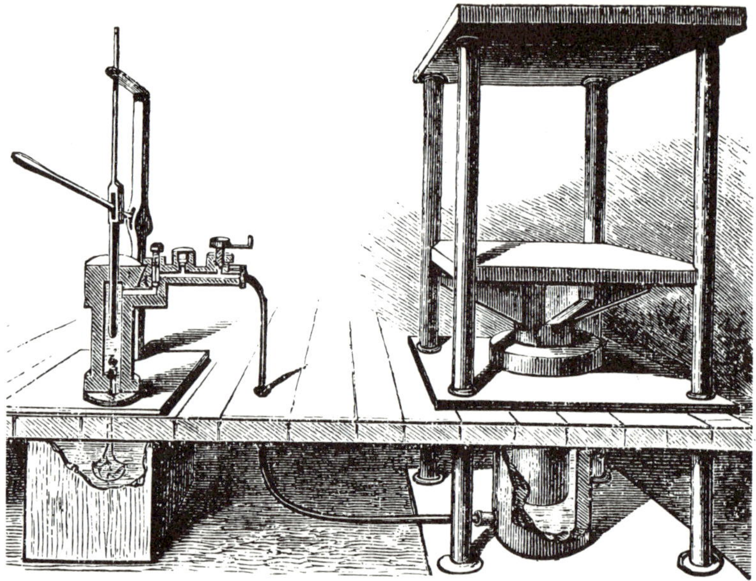

Fig. 80 Bramah's hydraulic press

used in this role to accomplish two major engineering feats. The first concerned the completion of the project to build the Britannia bridge over the Menai Straits. From a written account compiled by the resident engineer comes a vivid description of how Robert Stephenson raised four great tubular sections of the bridge into position. Each one was 144 m long and 400 tonnes in weight. Stephenson accomplished the task by means of two hydraulic jacks which were located on masonry supports.

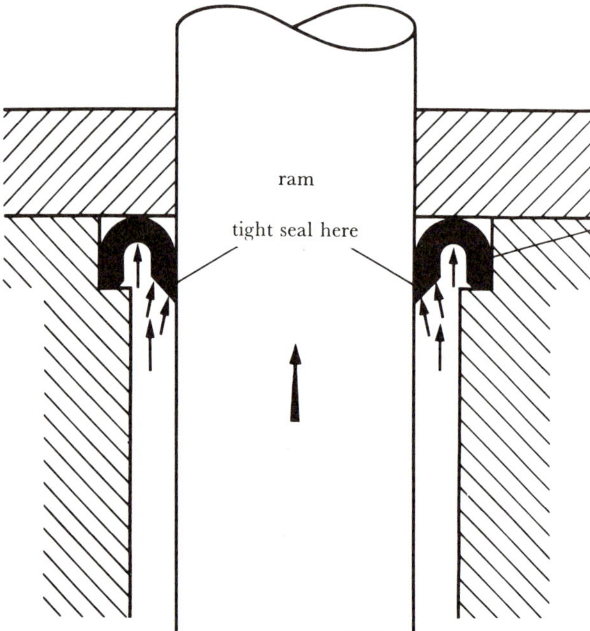

Fig. 81 Leather cup washer

The bridge was opened in 1850 and eight years later Brunel used hydraulic jacks to complete the launching of the Great Eastern sideways into the Thames. In 1860 a hydraulic multispindle drilling machine was designed which was capable of drilling eighty holes, 25 mm in diameter, with great

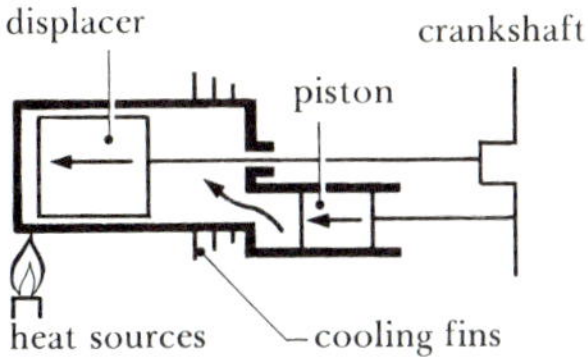

(1) Piston compresses air into cool space.

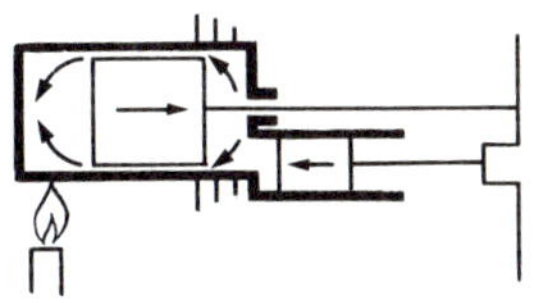

(2) Displacer transfers cool air into hot space.

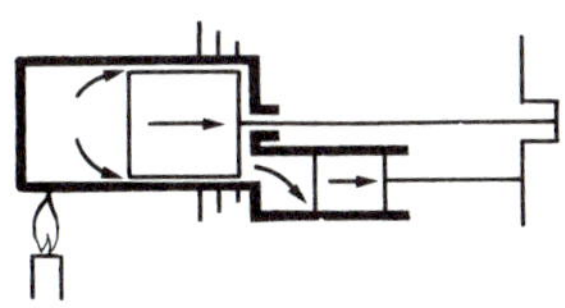

(3) Heated air expands pushing piston.

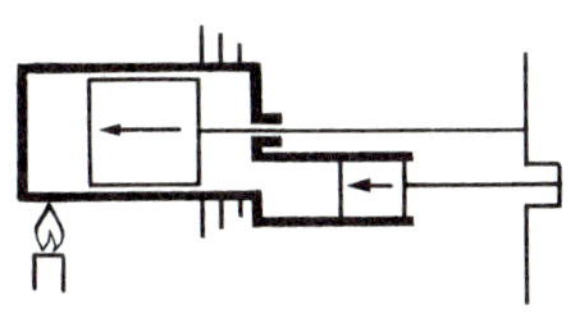

(4) Cycle repeats.

Fig. 82 The Hot-air engine: cycle of operations

accuracy. The first hydraulic crane built by the firm of Sir William Armstrong was demonstrated at Newcastle in 1846. Its adoption throughout Britain's docks conferred obvious advantages in cargo handling

The Hot-air Engine

A new source of power which enjoyed a limited degree of popularity in the early nineteenth century was the hot-air engine. Its invention is credited to Sir George Cayley and it could be used for such tasks as pumping water, driving fans, and operating lighthouse fog signals, operations where its economical performance gave it a big advantage over the steam engine.

Figures were quoted at the current values of that time which gave the costs of raising 1000 gallons of water through 150 feet as 1.3 pence for the steam engine and 0.75 pence for the hot-air engine.

The basic principle of operation involved the employment of the change of pressure caused by alternately heating and cooling an enclosed volume of air to drive a piston. One operating cycle is illustrated in fig. 82, which shows the two interconnected cylinders, a feature of the engine. The larger cylinder or displacer has a loosely fitted piston whose function is to move the air from the hot end to the cool end of the cylinder. The change in pressure of the air is communicated through a small aperture to the power cylinder which has a gas-tight piston.

Two types of hot-air engine were produced commercially; the closed cycle which we have described, and the open cycle which used a fresh supply of air for each operating stroke.

Sir George Cayley, better known for his work on the theory of aeronautics, constructed his first engine in 1807. In 1816 a Scotsman, Robert Stirling, designed an engine using the closed cycle. Other hot-air engines were produced by the Swede, Ericsson, and the Frenchman, Benier whose design was approved by Trinity House for use in lighthouses to operate fog signals.

The introduction of the electric motor and the light petrol engine in the closing years of the century made the hot-air engine obsolete. In recent years, however, there has been renewed interest in the hot-air engine. Scientists have shown that, theoretically, the engine is capable of higher efficiency than the diesel engine. Using modern materials in conjunction with our more extensive knowledge of heat transfer, the Dutch firm of Philips and General Motors of the U.S.A. have both developed hot-air engines with exciting possibilities. The problem of pollution caused by the waste products from the internal combustion engine has stimulated interest in the further development of the hot-air engine.

Internal Combustion Engines

THE GAS ENGINE

As early as the seventeenth century it had been established that a piston could be driven within a cylinder by using energy from an explosion. The very rapid combustion which resulted formed a large quantity of gas, and if the explosion

Fig. 83 Lenoir's gas engine

could be effected inside a closed cylinder these gases would expand and perform useful work in driving a piston. Papin had experimented unsuccessfully with gunpowder for this purpose before turning his attention to steam.

Impetus to the search for an engine working on these principles was given by the unreliability of the boilers in use at the time for producing high-pressure steam. Furthermore, a suitable fuel was now available in the form of coal gas, production of which had rapidly increased after its use for lighting had been demonstrated by William Murdock in the 1790s.

In 1860 Lenoir, a French engineer, designed and built the first successful gas engine (fig. 83). Operating on a mixture of coal gas and air, his engine was horizontal and double-acting. The mixture was ignited on alternate sides of the piston using an electrical spark, thereby giving two explosions for each revolution of the crank. Some 400 Lenoir gas engines were made in France between 1860 and 1865, with an output varying between 370 w and 2 kw. The engine was in many respects a derivation of the steam engine, and as no provision was made for the gas to be compressed before ignition occurred, the gas consumption of the machine was very high.

THE FOUR-STROKE CYCLE

In 1861 another Frenchman, Alphonse Beau de Rochas, described an ideal four-stroke cycle for internal combustion engines. The cycle comprised an induction stroke, a compression stroke, a power stroke and an exhaust stroke. The four-stroke cycle was first used by a German, Nicholas Otto, hence it is commonly referred to today as the Otto four-stroke cycle, and is, of course, used in the majority of modern internal combustion engines. The

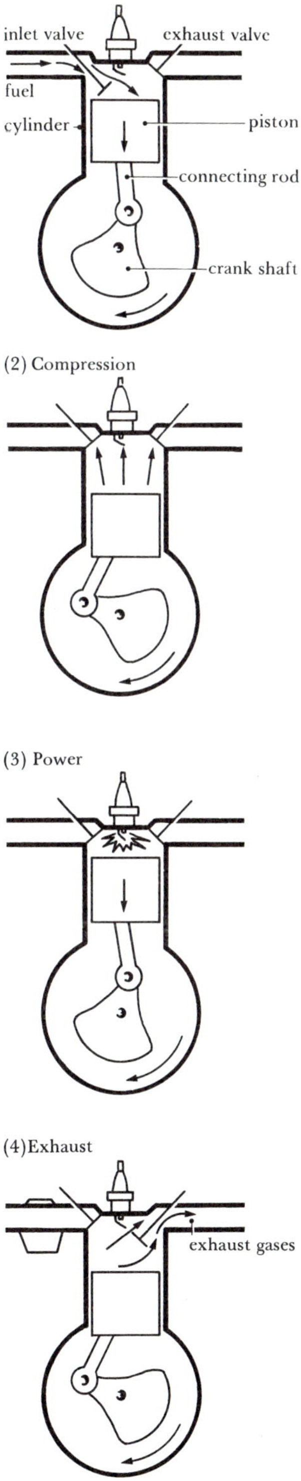

Fig. 84 The Otto four-stroke cycle

Otto gas engine (fig. 85), introduced in 1876, established a great reputation for silence and reliability, and many thousands were produced both in Germany and under licence in other countries.

A single-cylinder engine operating on the Otto four-stroke cycle had one power stroke for every two revolutions of the crankshaft. To keep the engine turning smoothly throughout the cycle it was essential to fit a heavy flywheel to the end of the crankshaft. In multicylinder engines, which were not developed until the twentieth century, power was applied to the crankshaft at intervals throughout its revolution and there was less dependence on a flywheel, although it was still incorporated.

It will be remembered that the cylinder of the steam engine was provided with a steam jacket to raise the temperature of the cylinder walls and so prevent heat losses. In the internal combustion engine, however, the great heat generated by the explosion of the gases is absorbed by the cylinder walls which must be provided with a water jacket to enable the heat to be dissipated via a radiator.

Early gas engines relied on town gas supplies, but towards the end of the century large gas engines were operating on gas from coke ovens. In steel works the waste gases from blast furnaces were used to power engines which drove electric generators. Many such engines are still performing useful work today.

Fig. 85 Otto gas engine

OIL

The great disadvantage of the gas engine lay in its dependence on an adequate gas supply. This caused inventors, who were looking to the internal combustion engine as a hopeful source of motive power, to explore the possibilities of using a liquid fuel which could be easily transported and stored. The fuel with which they hoped to solve this problem was natural oil which was becoming available in increasing quantities as drilling methods improved. Natural oils were broken down by distillation into oils having different properties. The highest fraction, i.e. the first portion to be vapourised, is the one we know as petrol or gasoline, after which came kerosene, or paraffin, and then the medium and heavy oils. Development of the petrol engine and the oil engine progressed simultaneously during the last quarter of the nineteenth century.

THE PETROL ENGINE

The first practical petrol engine appeared in 1885, designed independently by the Germans, Gottlieb Daimler and Karl Benz. Both inventors had to overcome two major problems before achieving success. The first was concerned with the induction of the correct mixture of petrol vapour and air into the cylinder at the appropriate point in the four-stroke cycle. The second involved designing a suitable method of igniting the mixture in the cylinder after it had been compressed by the piston.

Daimler, an experienced gas engine designer who had worked for Otto and Langen, at first used a method which entailed sucking warm air through a layer of petrol to solve the induction problem. By 1893, however, his partner, Wilhelm Maybach, had invented the jet carburettor, which incorporated a float chamber (fig. 86), the principles of which are embodied in many modern carburettors. Ignition for the first Daimler engine was provided by an externally heated tube which passed into the cylinder. Daimler engines ran at the high speed of 900 r.p.m., compared with the 200-300 r.p.m. of the gas engine. Later, in 1885, he installed a second model of this engine in a bicycle frame, thereby creating the first motor cycle.

The first automobile was produced in 1885 by Karl Benz, and it took the

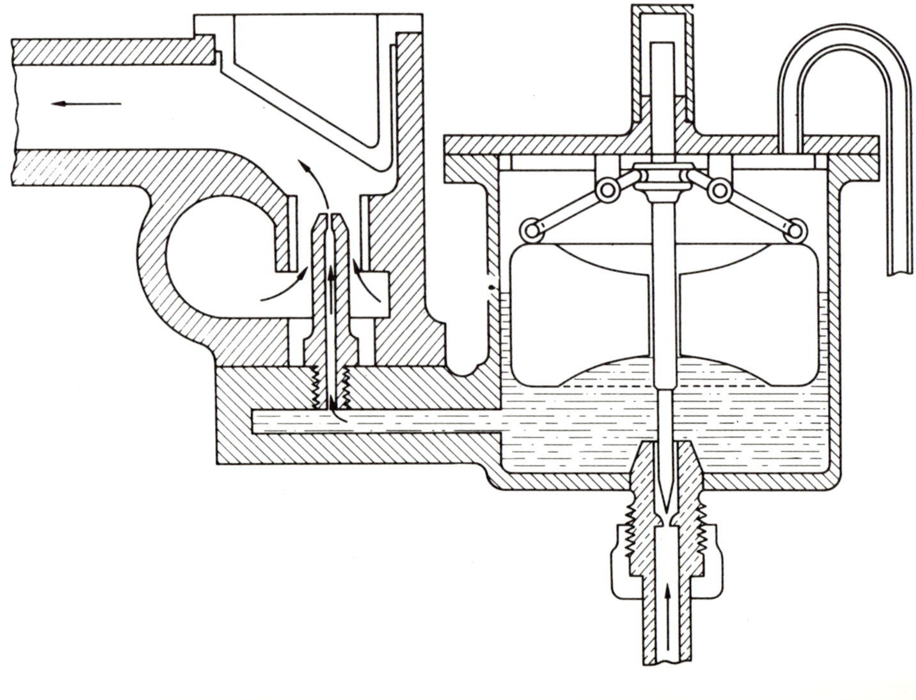

Fig. 86 An early float feed carburettor

form of a tricycle driven by a petrol engine of his own design. Benz used a surface carburettor to supply the correct mixture of petrol vapour and air which operated by sucking air across petrol arranged in a series of shallow trays. More important was the design of the ignition system which embodied all the essentials of the one used in a modern car.

Using an induction coil and contact breaker, a spark was made to jump across a gap between two insulated wires incorporated in a plug which protruded into the cylinder. Benz's engine was fitted with a cooling system using a radiator and he made use of the differential gear to drive the axle. These engines were designed to run on benzene, a derivative of coal-tar, as supplies of petrol of adequate quality were difficult to obtain at that time. The carburettor in the Benz engine was provided with a shutter or choke which controlled the inflow of air, and thus acted as a speed regulator.

F. W. LANCHESTER (1868-1946)

Although the mainspring of early petrol engine development was French and German, a noteworthy contribution to the design of petrol engines for the motor car was made by the Englishman, F. W. Lanchester. Working in Birmingham he produced his first motor car in 1895 which incorporated many original ideas. His wick carburettor was a reliable alternative to the Maybach jet carburettor which suffered from blocked jets caused by the impurities in the fuel then available.

Lanchester's engine had two horizontally opposed cylinders (fig. 87). Each piston had two connecting rods to drive a pair of contra-rotating crankshafts. The design obviated excessive stresses in the crankshafts and gave better balance. Lanchester adapted his car to suit the new form of propulsion, and in so doing he differed from many contemporary car makers who were content to install engines in bodies which derived directly from the horse drawn carriage.

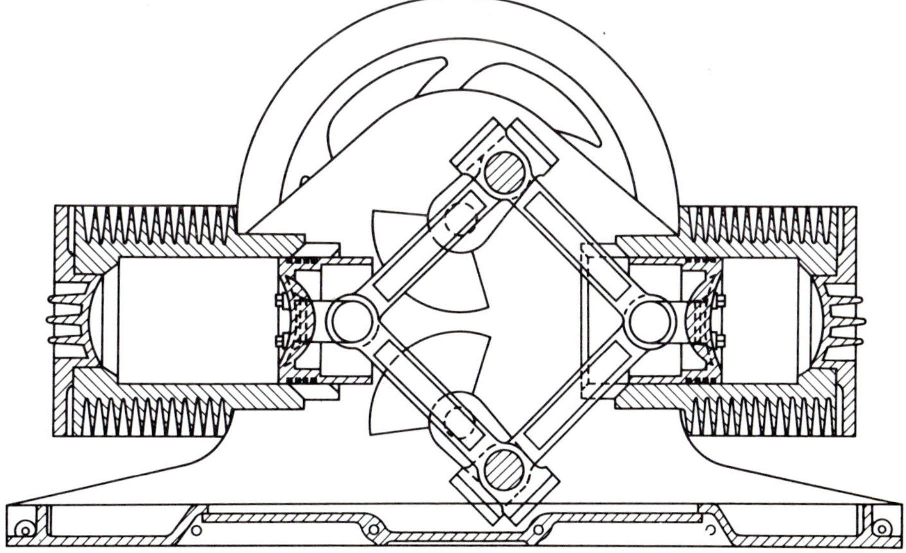

Fig. 87 The Lanchester engine

In his second car, completed in 1898, he used an epicyclic gear box and replaced the chain drive of the first car by a propellor shaft fitted with universal joints. The drive to the back axle was through a worm wheel. To produce his cars efficiently Lanchester designed special machines and innovated new production techniques, including a system of tolerances which facilitated the interchangeability of parts.

THE OIL ENGINE

Petrol vaporises readily when exposed to the air, but if the heavier fractions of natural oil are to be used as a fuel for internal combustion engines, it is essential that the oil should be converted into a fine spray or mist.

One of the first oil engines to achieve commercial success was built by the Priestman brothers of Hull. This was a single cylinder four-stroke cycle engine which developed about eleven horse power using paraffin as a fuel. The paraffin was injected into a vessel heated by the exhaust gases called a 'vaporiser' where it immediately became gaseous. On the induction stroke of the piston the gas was mixed with air and drawn into the cylinder. After compression the mixture was ignited and the power stroke followed.

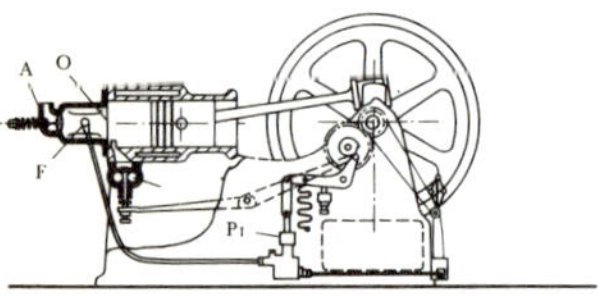

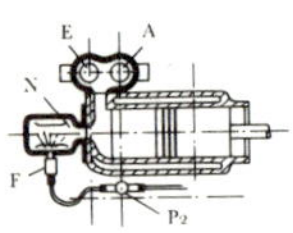

Fig. 88 The Hornsby-Stuart oil engine

The Hornsby-Stuart oil engine (fig. 88) which appeared in 1890 heralded a significant improvement in that it dispensed with the external vaporiser and ignition apparatus. Designed by Ackroyd Stuart and manufactured by Ruston and Hornsby of Lincoln, it incorporated a device known as a 'hot bulb' at the top of the cylinder. The bulb received preliminary heating from a blow-lamp, and as the engine was turned by hand, air was sucked into the cylinder through a valve, while at the same time oil was pumped into the 'hot bulb' where it vaporised. On the compression stroke air was forced into the hot bulb and mixed with the oil vapour. At the end of the compression stroke the compressed mixture was ignited by the heat of the bulb and the expanding gases passed into the cylinder to produce the power stroke. After the first few strokes external heating of the bulb was dispensed with as it retained sufficient heat from the combustion of the fuel to maintain continuous running.

THE DIESEL ENGINE

When air is subjected to sudden pressure in a confined space considerable heat is generated. A toy device popular in the nineteenth century demonstrated this principle very clearly. Tinder fixed to a piston was ignited when the piston was pushed quickly down a closed tube. If an inflammable vapour could be injected into the cylinder at the completion of the piston stroke, the heat generated by the compression of air would be sufficient to ignite it. This was the idea which when incorporated in the internal combustion engine was to become known as compression ignition.

In 1892 a German, Rudolph Diesel, designed a heavy-oil engine to work on a compression-ignition system, and in the following year published a book in which he described the principles involved in its design. His engine built in 1895 achieved compression far higher than was required for self-ignition in order to obtain the greatest possible fuel efficiency. The injection of the fuel was accomplished using air compressed to a much greater pressure than that inside the cylinder. The injected fuel ignited the moment it entered the cylinder and burned smoothly, driving the piston downwards as the gases expanded. No explosion occurred in the cylinder of a Diesel engine.

To withstand the high compression the engine was of robust design and required ancillary equipment to compress the air used for injection. The engine achieved a high efficiency with a low fuel consumption, and it was capable of using crude oil which previously had been impossible to sell. In later engines an air-compressor driven by the engine itself was incorporated to charge a steel cylinder up to pressures of 70 kg/cm^2.

Diesel used the four-stroke cycle in his engines, but later engines of the diesel type were built to operate on the two-stroke cycle, first used by Sir Dugald Clark in a gas engine which he built in 1878. In the two-stroke cycle there is a power stroke for every revolution of the crank, exhaust and induction taking place simultaneously at the end of the power stroke.

The development of the internal combustion engine progressed with such rapidity that by the end of the nineteenth century many varieties of gas, oil, and petrol engines were available, using either the four-stroke or the two-stroke cycle.

LUBRICATION

In the course of the nineteenth century every aspect of mechanical engineering displayed a greater awareness of lubrication and its advantages. An informed French traveller visiting the Cyfarthfa Ironworks in Wales in 1823 extolled the benefits of using high-quality Russian fat as a lubricant.

Following the U.S. oil boom of the mid-1850s there was a switch to mineral oils which offered a wide range of possibilities. Engineers were soon able to distinguish the important properties of a lubricant; its viscosity, oiliness and its propensity to gum or freeze. Ways were found to lubricate fast-moving parts without stopping motion. Rubbing faces of a machine were placed in a standing bath of oil or a saturated pad was fixed so that the parts moved against it. By the 1890s forced lubrication allowed for really high speeds with a minimum of engine wear.

Power from Electricity

Until about 1850 electricity was regarded as a scientific curiosity rather than a potential source of power. Nevertheless, the early laboratory experiments were of vital importance in gaining an understanding of the behaviour of electricity.

In 1786 Galvani carried out his famous experiments with frogs' legs, observing that they twitched when the nerve was touched by a rod made of copper and zinc. Galvani ascribed the movement of the legs to what he called 'animal electricity', but another Italian physicist, Volta, was able to draw the correct conclusion that an electric current had been produced by two dissimilar metals being in contact. From this conclusion he went on to construct the Voltaic Pile, which was a primary electric battery, so that in 1800 a source of continuous electric current was available.

An important contribution was made by a Danish professor, Hans Oersted, who in 1820 discovered the existence of electromagnetism. He found that when he passed a current along a straight wire a compass needle held above it was deflected. Within a few weeks the French scientist, Ampère, had established that a coil of wire carrying an electric current acted in the same manner as a magnet and that two such coils displayed the same powers of attraction and repulsion as a pair of magnets.

In England the phenomenon of electromagnetism occupied the attention of Michael Faraday for a number of years. By 1831 he had established that an electric current could be induced in a conductor if it were passed through the field of a permanent magnet. Inspired by the idea of producing electricity by mechanical means he demonstrated, in 1831, a machine which would do this. Faraday caused a copper disc some 30 cm in diameter to rotate between the poles of a permanent horseshoe magnet. A wire was taken from the rotating shaft and another was connected to a brush in contact with the outer portion of the disc. When the free ends of the wire were connected to a moving coil galvanometer, displacement of the galvanometer needle showed that a current was flowing in the wire.

In April of the following year Faraday published a paper giving the results of his work on induction. He had wound two separate coils of wire on opposite sides of a closed iron ring and then connected the ends of one coil to a battery and the ends of the other coil to a galvanometer. As the circuit to the battery was completed the galvanometer needle moved in one direction and then returned to zero. When the circuit was broken a similar movement of the needle, but in the opposite direction, was observed. Later in the same year an American, Joseph Henry, made a similar discovery.

With these two historical experiments Faraday demonstrated the basic principles of the electric generator and the transformer. Although it was many years before these principles were embodied in commercial machinery his experiments inspired increasing activity amongst fellow scientists.

In 1832 a Frenchman, Pixii, designed a hand-operated generator producing electricity in a fixed coil of wire by rotating beneath it a horseshoe magnet. A simple commutator was incorporated in the machine in order to provide direct current. Later inventors established the pattern of rotating a number of rectangular coils of wire (an armature) in the field created by fixed permanent magnets, to provide electrical current. As we have seen, the current induced in these coils reverses every 180 degrees of rotation to produce what is called an alternating current. To produce a direct current, considered more desirable by early experimenters, the ends of the coils had to be fitted to a commutator which counteracted this reversal.

Wheatstone substituted electromagnets for permanent magnets in 1846, and by 1866 self-excitation of the electromagnets was in use in which the current to excite them was supplied from the armature. Improved armature design followed from the work of Pacinotti who introduced a ring armature and improved commutation in 1860. The ring armature, which solved the problem of connecting in series the conductors of a multipolar dynamo, was used by Gramme whose generators designed to be driven by steam engines were in widespread use during the 1870s.

The ring type armature was eventually replaced by the drum type winding of continuous wire with which we are familiar. The drum armature, the invention of Werner von Siemens in 1856, was improved by Alterneck in 1871, and later by Ferranti.

Most of the early work in electrical power was concerned with the production of generators, although Davenport in America was using electrical motors to perform drilling and woodturning operations as early as 1837. At this early stage, however, the expense of providing electricity from batteries was prohibitive.

The possibility of converting mechanical energy to electrical energy for transmission by cables to an electric motor which would re-convert the electrical energy into mechanical energy was demonstrated by Gramme. In Vienna in 1873 he used the same type of machine to act as a generator and a motor. This, and a later demonstration by Siemens in 1878, paved the way for using electrical power to operate prime movers.

The invention of the carbon filament lamp which was perfected by Edison and Swan in 1880 stimulated the demand for electrical power, and the first generating station was built by Edison in Pearl Street, New York, in 1882. The early generating plants supplied direct current from generators driven by reciprocating steam engines. By the 1890s the advantages of an alternating current supply were becoming better understood, transformers were utilised to step-up the voltage for transmission over a distance, and then to step-down the voltage to that required by the consumer.

The generators, or, more correctly, alternators, for supplying alternating current were driven by steam turbines in the later years of the century, and in 1903 a turbo-alternator generating 5000 kw was working in Chicago. For a large part of the nineteenth century, power for factory machines was generally supplied from a large stationary steam engine located in an engine house outside the workshop. Power for the individual machinery was taken from overhead shafting using leather belts and pulleys. The substitution of a large electric motor for the steam engine had definite advantages. In particular the motor could be located inside the workshop so that large outbuildings were no longer necessary. Even more efficient was the practice, which spread from the United States in the years immediately before 1900, of providing each machine with its own electric motor.

In 1883 the Siemens brothers and John Hopkinson built the first electric traction system in Great Britain for the Portrush railway. Electricity was generated by a water turbine, and the current at 240 volts was picked up from a third rail. In 1890 the South London line was opened, using electric locomotives designed by Siemens. In the U.S.A. numerous installations involving electric tramways operated by power taken from overhead cables or live rails, were built in the 1890s. Among the many inventors and designers involved the names of F. J. Sprague and C. J. van Depoele are noteworthy.

Additional Reading

T. K. DERRY and T. I. WILLIAMS, *A Short History of Technology,* O.U.P. 1960.
Engineering Heritage, 2 vols, Institution of Mechanical Engineers 1963, 1965.

Chapter 7
Measurement

Standards of Length

From the moment that civilised man started to divide up the available land and construct his dwellings a system of length measurement based on commonly acceptable standards was essential. It would be natural for him to use the parts of his body for this purpose. The feet, the arms, the stride, and the span, although varying between individuals, could be adopted as units of length where accuracy was not important. As building techniques became more sophisticated in the early civilisations of Egypt, Sumeria, and Babylon, it became essential to standardise these units of length to eliminate the human variation. This was achieved by marking the required unit of length on some hard wearing material. The length or standard was then locked away in a safe place, and copies of the standard were made and checked as required.

In this way the cubit, originally the distance from the elbow to the tips of the fingers, was standardised by the Egyptians as the Royal Cubit. Its length was equivalent to 20.6 inches (52.3 cm). Similarly, the Sumerians, Assyrians, and Greeks all fixed their own standard length for the cubit.

Larger units of length were used by the Greeks. The stadion, for example, was a distance of 600 'feet', although the length of the Greek 'foot' is not known exactly. In the second century B.C. Eratosthenes calculated the dimensions of the earth using the stadion as his unit of length. The Romans divided their 'mile', which was based on 1000 double paces, into eight divisions called 'stadia'. The Roman mile was the equivalent of about 1665 yards and formed the basis of later distances of one mile.

Laws governing measures and weights were in existence in this country before the Norman Conquest although they were observed only loosely. The Anglo-Saxon yard was 39.6 inches, but in 1101 it was shortened by Henry I who decreed that the standard yard should be the distance from his nose to the end of his thumb when his arm was stretched out sideways. A law of Edward I in 1305 laid down that 'three barley corns, round and dry, made one inch, twelve inches one foot, and three feet one ulna (yard)'.

In the sixteenth century one authority defined the length measurement known as the rod in the following terms 'Stand at the door of the church on Sunday, and bid sixteen men to stop, tall ones and short ones, as they happen to pass out when service is finished; then make them put their left foot one behind the other and the length thus observed shall be the right and lawful rod to measure and survey land with, and the sixteenth part of it shall be the right and lawful foot.' As the rod had previously been fixed as a length of $5\frac{1}{2}$ ulna, that is $16\frac{1}{2}$ feet, this definition would suggest that there was still an element of local variation in the determination of standards.

Nevertheless the standard yard decreed by Edward I formed a surprisingly steady basis for future linear measurements. A standard yard, based on that of Edward I, was produced to the order of Henry VII in 1497, and this in turn was replaced by one introduced in 1588, in the reign of Elizabeth I. The discrepancy between this standard and our present United Kingdom primary standard yard is only 0·01 inches. The total variation in the length of the standard yard between 1305 and the present day is as little as 0.04 inches.

THE UNITED KINGDOM PRIMARY STANDARD OF THE YARD

The present standard yard was manufactured in 1845 and was made legal by statute in 1856. It consists of a one inch square bar of bronze (16 copper,

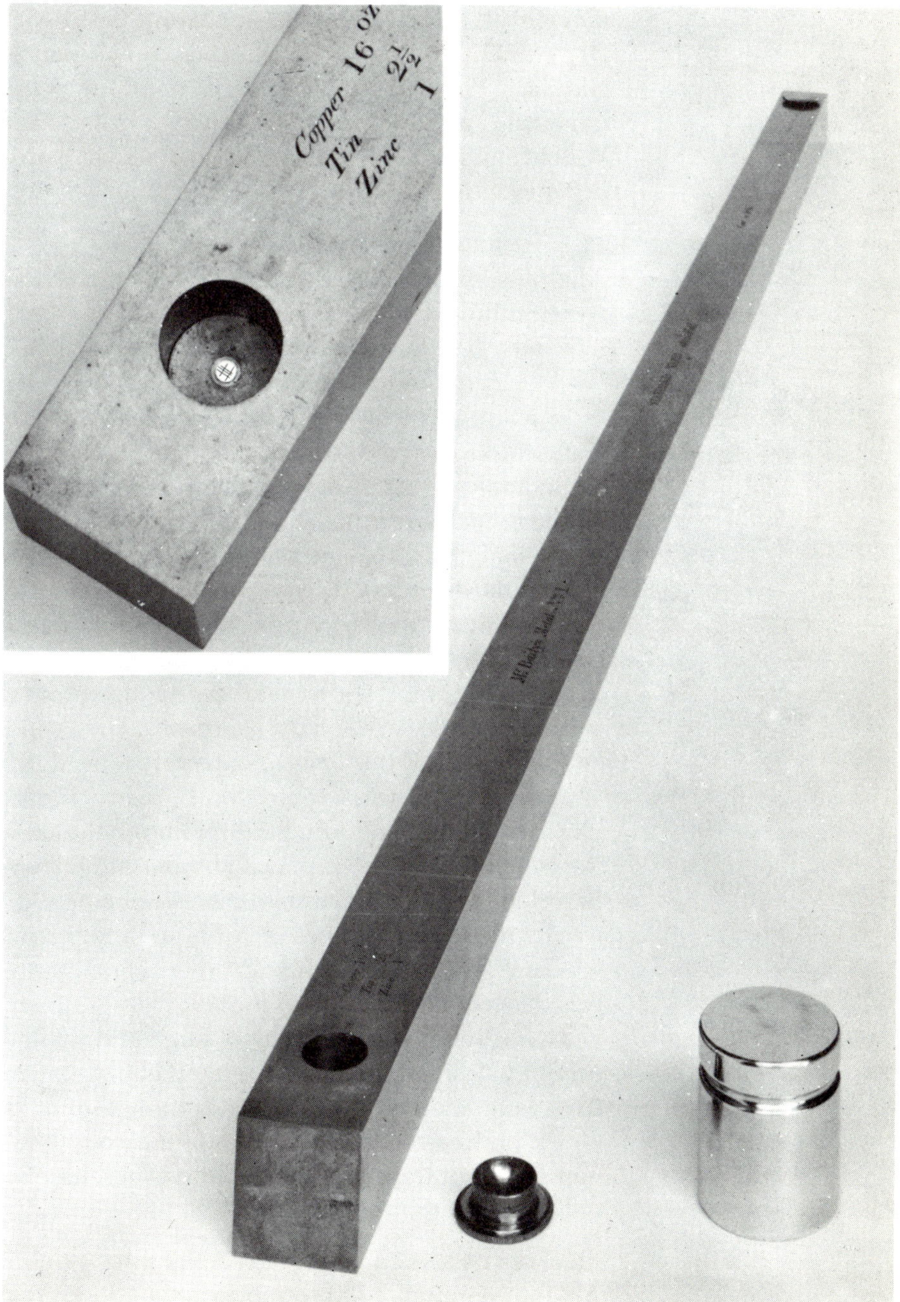

Fig. 89 U.K. Primary Standards of the yard and the pound

$2\frac{1}{2}$ tin, and 1 zinc) thirty-eight inches long. One inch from each end of the bar a hole approximately half an inch in diameter is drilled to half the depth of the bar. At the bottom of these two holes a gold plug is inserted so that its surface coincides with the axis of the bar. On the face of each plug two lines are engraved parallel to the axis (fig. 89). The yard is defined by the Weights and Measures Act of 1963 as the distance between the middle of the three transverse lines at each end when the temperature is 62°F and the pressure of the atmosphere supports thirty inches of mercury.

The U.K. Primary Standard Yard (formerly the Imperial Standard Yard) is preserved by the Department of Trade, and Authorised Copies are held by the Royal Society, the Royal Mint, Greenwich Observatory, and the Department of Trade and a further copy is deposited at Westminster. The National Physical Laboratory is responsible for the comparison of the standards, at intervals determined by the Department of Trade. The Weights and Measures Act 1963 defines the yard in terms of the metre as being exactly 0·9144 m.

THE METRIC SYSTEM

The possibilities of introducing a logical system of weights and measures were examined as early as the sixteenth century. In the latter part of the seventeenth century a group of scientists which included Sir Christopher Wren, Picard, and Huygens suggested a length standard based on the length of a pendulum which would beat half-seconds. In 1670 a French priest, Gabriel Mouton, proposed a system based on the length of the arc subtended by an angle of one minute on a great circle of the earth.

The French government became interested in a new system to replace their traditional measurements which were so haphazard that some twelve different standards of length were in use throughout the country. In 1790, the second year of the Revolution, a committee was appointed to devise a new system of weights and measures and determine the standards.

Initially it was decided that the standard unit of length, to be called the metre, was to be based on the length of a one-second pendulum. The scientific determination of this length, however, proved difficult, and in 1792 a provisional standard was adopted for the metre. Finally, the standard metre was taken as one ten-millionth part of a line on the earth's surface stretching from the North Pole to the Equator and passing through Dunkirk in France and Barcelona in Spain. The survey of the line between Dunkirk and Barcelona was hampered by war between France and Spain, and the final calculation was later found to be slightly in error. The distance accepted as the standard was the equivalent of 39.37 inches.

To facilitate the measuring of large and small distances other units were taken by decreasing and enlarging the metre by multiples of ten. Latin prefixes were adopted for the smaller units—deci, centi-, and milli-, and the Greek prefixes—deka, hecto, and kilo were used for the larger units. The metric system was not accepted into general use in France until it was made obligatory in 1840.

THE INTERNATIONAL PROTOTYPE METRE

An International Convention in 1875 proposed the construction of a new standard metre—to be called the International Prototype Metre. This was constructed from an alloy of platinum and iridium and took the form of a bar of cruciform section. The metre was defined as the distance between two lines engraved on the neutral plane of this standard when the temperature of the bar was nought degrees Centigrade. Copies of this standard are held by countries which were signatories to the Convention of 1875. The British copy, now called the U.K. primary standard of the metre is held by the Department of Trade, and is checked against the International Prototype, which is preserved at Sèvres near Paris, at determined intervals.

STANDARDS OF WEIGHT

Weighing by means of the balance dates from about 4000 B.C., and apparently its earliest use was for the weighing of gold. The balance with two pans, one to hold the weights and the other to hold the commodity, was in use in the civilisations of the western Mediterranean by 2000 B.C. The shekel was the standard weight used, varying between seven and fourteen modern grammes, larger units were the mina and the talent which was approximately equivalent to a hundredweight. The steelyard, the type of balance still to be seen in butchers' shops, was introduced by the Romans.

In 1266 a statute of Henry III made legal a table of weights based on a grain of wheat taken from the middle of the ear. Thirty-two such grains were to be the weight of one silver penny, twenty pennyweights made one ounce, and there were twelve ounces to the pound. The pound was also to be the weight of twenty shillings. It will be seen that this system of weights bore a direct relationship to the monetary system of the period, and in fact coins were used as weights.

The avoirdupois system with its sixteen ounces to the pound was adopted early in the fourteenth century, and was used at first only for weighing bulky commodities. The two systems continued in use side by side, with the earlier system becoming known as Troy weight. The Weights and Measures Act of 1878 abolished the Troy pound, and the Troy ounce was reserved for use in the gem and precious metal trades.

In the reign of Elizabeth I a new set of standard weights was prepared, and the hundredweight was fixed at 112 pounds avoirdupois, with twenty hundredweights to the ton. The Pilgrim Fathers established the avoirdupois system in America, but the hundredweight became fixed at 100 pounds giving a ton of 2000 pounds. Thus the U.S.A. ton became known as the 'short' ton.

Although we have used the word weight so far in its commonly accepted sense it is more correct for engineers and scientists to refer to a mass of one pound. The weight of one pound being the force exerted on a one pound mass by the earth's gravitational field. This is more usually stated as that force which, when applied to a mass of one pound, produces an acceleration of 32.2 feet per second per second.

Weights, volumes, areas, and liquid measures, were all rationalised when the metric system was being created in France. A logical system was designed in which these quantities were expressed in terms of the metre. The standard unit of mass, the gramme, was taken to be the mass of one cubic centimetre of water at a temperature of 4°C. Larger and smaller units in multiples of ten were named with the same prefixes as those used for the linear units.

THE UNITED KINGDOM PRIMARY STANDARD OF THE POUND
The U.K. Primary Standard Pound, formerly the Imperial Standard Pound, was produced in 1856 and is defined by the Weights and Measures Act of 1963. It is a cylindrical solid of platinum with a small groove cut around its surface so that it may be lifted with an ivory fork. The arrangements for the disposal and testing of the Authorised Copies are the same as those for the Standard Yard. The Weights and Measures Act 1963 defines the pound in terms of the kilogram as being exactly 0.45359237 kg.

THE INTERNATIONAL PROTOTYPE KILOGRAMME
This was established in 1889 and consists of a cylindrical solid of an alloy of platinum and iridium with its diameter equal to its height. The arrangements for copies are the same as those for the Prototype Metre.

Measuring Tools and Instruments

The measuring tools of the medieval millwright were the square, calipers, dividers or compasses, and the ruler. This was equipment that had been used by craftsmen from antiquity, and to which little addition was made until the nineteenth century.

The progress in the skills of instrument making took place through the demands of astronomers and navigators in search of more accurate devices to aid their computations. In the fifteenth century instrument making was established in Europe with techniques acquired from the Arabian civilisation which in turn was carrying on the traditions of the famous Alexandrian astronomers of the first century B.C. The craft of instrument making was brought to England in the early sixteenth century by German immigrants.

As well as making the familiar compasses, squares, and rulers, they also produced astrolabes, cross-staffs, and quadrants. The astrolabe was used by navigators to determine latitude, while the cross-staff measured angular distances. Quadrants were employed by gunners and surveyors to obtain angles of elevation and depression.

Fig. 90 Use of the quadrant 1547

Engraving on brass was a particular accomplishment of the instrument makers, and the introduction of the telescope in the seventeenth century extended their skills to the production of optical lenses. During the eighteenth century their skills were in constant demand to provide scientists with recently discovered instruments such as the air pump, the thermometer, and the barometer.

Accuracy was increasingly a concern of the astronomers whose work commanded attention because it was of direct use in navigation. Both the micrometer and the Vernier scale, which were to become essential items in the equipment of the engineering craftsmen, were invented for use in astronomy.

PRECISION ENGINEERING IN THE NINETEENTH CENTURY

The rapid development of machine tools in the first half of the nineteenth century enabled the engineering craftsmen to work accurately to much finer limits than had hitherto been possible. In order to construct the machine tools it was essential to establish entirely new and more demanding standards of accuracy. Of paramount importance was the ability to produce a plane surface, and the practice of checking work for flatness against a master surface started in the workshops of Henry Maudslay.

Maudslay is said to have been the originator of the practice of making surface plates in threes. These were rubbed one against the other, after being coated with coloured dye. High spots were revealed where the dye was rubbed away, and these were removed by hand scraping. If only two plates had been used, it would have been possible for them to be rubbed together revealing no high spots and yet not having true plane surfaces.

Sir Joseph Whitworth carried on the exacting standards of the Maudslay workshop and campaigned for higher standards of accuracy throughout the machine tool industry. He emphasised the need for using true plane surfaces as a basis for accurate work, and strongly condemned the contemporary practice of producing surface plates by grinding the faces together using emery powder. This method only reproduced any existing flaws in the surfaces.

The practice of using interchangeable parts was pioneered by French gunsmiths at the end of the eighteenth century. By the mid-nineteenth century, it had moved over the Atlantic where the so-called American System became the forerunner of mass production. The success of the system depended upon the ability to manufacture parts within specified limits in order to permit interchangeability. To this end the use of drilling and filing jigs which had also originated in Europe became standard practice in the workshops of Whitney, North and Colt. Gauges were also used to check that the parts were within acceptable limits of accuracy.

THE VERNIER CALIPER

The Vernier scale was invented by Paul Vernier about 1630. He used it for the accurate measurement of angles in astronomy. It was adapted for use in other scientific instruments, both for angular and linear measurements. When a standard of accuracy in the engineering industry was no longer fulfilled by the ruler it was logical that the Vernier scale should be incorporated into a measuring tool.

The first Vernier calipers were manufactured by the American firm of Brown and Sharpe in the 1850s. The accuracy of these instruments was enhanced by the use of a machine invented by Brown in 1850 for the accurate graduation of rulers. From this time onwards a tool capable of measurements accurate to a one-thousandth part of an inch was available at a price within reach of every craftsman.

THE MICROMETER

The invention of the micrometer, a tool capable of accurate measurement and based on the screwthread, is credited to William Gascoigne who was killed at the Battle of Marston Moor in 1644. Gascoigne had developed his micrometer for use in astronomy.

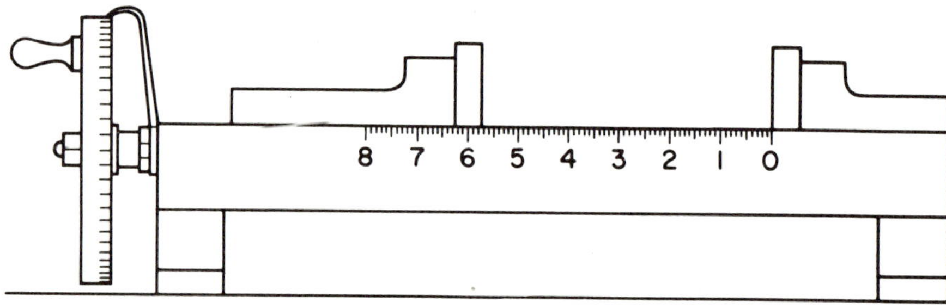

Fig. 91 Maudslay's micrometer 1829

Probably the first engineer to make use of the principle was Maudslay. As the accuracy of the micrometer is dependent on the accuracy of the screwthread, Maudslay was well placed to produce his famous bench micrometer (fig. 91). He referred to this instrument as the Lord Chancellor, basing its name on the pun 'that it was the final court of appeal'.

In 1851 Joseph Whitworth produced a measuring device based on the micrometer, which he exhibited at the Crystal Palace. The instrument was capable of measuring to an accuracy of one-millionth of an inch, and was used to produce accurate end gauges.

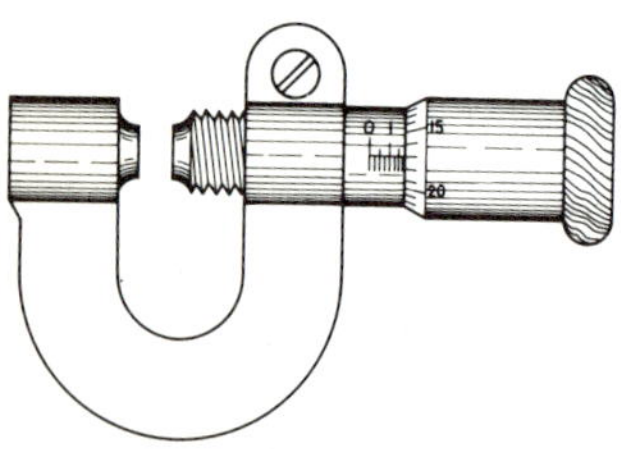

Fig. 92 The Brown and Sharpe micrometer

The small pocket micrometer with which we are familiar was first produced commercially by the firm of Brown and Sharpe. The Brown and Sharpe model incorporated the best features of two existing micrometers, one the work of an American, S. R. Willmot, and the other of a Frenchman named Palmer. The original Brown and Sharpe micrometer (fig. 92) included a clamping device for locking the spindle and a means of adjusting the anvil to compensate for wear.

Standardisation in Engineering

END STANDARDS AND LINE STANDARDS

Standard lengths such as the Standard Yard and Prototype Metre are called line standards. The length being defined as the distance between two lines. End standards are measured between two flat parallel surfaces. The micrometer and the Vernier caliper are end measuring instruments.

Making accurate copies of line standards with equipment available in the first half of the nineteenth century was extremely difficult. In any case the accuracy of measurements taken from graduated copies of line standards such as rulers would vary with the skill of the craftsman.

To achieve the accuracy required to make parts interchangeable under the American System, the use of end standards was adopted. A skilled craftsman produced an accurate prototype of the article to be mass-produced. Sets of gauges were then made from the prototype so that the dimensions of the component parts could be checked. It will be seen that although accuracy within a particular firm could be controlled by this method, each firm was creating its own standards and there was little chance of national uniformity.

In America concern about the variation in standard gauges caused the firm of Pratt and Whitney to investigate the situation. They found that their own standard gauges were at variance with those of other firms and with official standards. A professor of Harvard University, William Rogers, backed by Pratt and Whitney, undertook the determination of precise standards. Working from the Imperial Standard Yard and the International Prototype Metre and comparing them with the U.S. Standard Yard, he constructed a standard for Pratt and Whitney. To achieve an accurate comparison and subdivision of this standard, Rogers, in collaboration with George Bond, developed the comparator. This instrument made use of a travelling microscope with a micrometer eyepiece. Apart from its value in comparing line standards it could also be used for checking end gauges against line standards. Thus it was now possible for accurate and interchangeable sets of standard gauges to be used throughout the country.

SCREWTHREADS

While a number of larger firms such as Maudslay, Sons and Field, had their own standard range of screwthreads, the situation throughout Britain and America was chaotic. It was usually necessary to send to the manufacturer if a nut was required to fit a bolt on a particular machine. Concern about this state of affairs prompted Sir Joseph Whitworth to investigate screwthread practice in Britain, and in 1841 he proposed a standard screwthread based on those used by leading manufacturers. This thread, later to become familiar as the British Standard Whitworth, was of vee-form with a thread angle of fifty-five degrees. The top and bottom of the threads were rounded off to minimise damage which might occur to the threads and to the taps and dies which produced them. Whitworth also prepared a table giving the number of threads per inch for different bolt diameters. By 1860 the Whitworth thread had been adopted in the heavy engineering industry throughout the country.

In America the standardisation of screwthreads was undertaken by William Sellers who, in 1864, proposed the adoption of a vee-thread having a sixty degree angle with a flat crest and root. This thread was quickly adopted by American industry and was in general use by 1870. The Sellers thread form was later accepted by most of the continental countries.

The fact that the two foremost industrial nations of the nineteenth century used different screwthread standards caused expense and inconvenience as world trade developed. During the two world wars the difficulties were extreme and Britain's final acceptance of metric standards in 1972, however agonising, must be seen as a step of great significance.

WIRE AND SHEET METAL GAUGES
The Brown and Sharpe Wire Gauge, proposed by Lucian Sharpe, was adopted by American industry in 1857 and was later used for gauging sheet metal. In Britain a proliferation of wire and sheet metal gauges were in use throughout the country. These were rationalised by the introduction of the Imperial Standard Wire Gauge as the official gauge for measuring the diameter of wire and the thickness of non-ferrous sheets. The Birmingham Gauge was recognised by an Order in Council of 1914 as the official gauge for iron and steel sheets. At the end of the nineteenth century Britain lagged behind the United States and Germany in many aspects of engineering standardisation. Pressure from leading British technical institutes, concerned over the country's competitive position, resulted in the formation of the Engineering Standards Committee (1901) later to become the British Standards Institution.

Engineering Drawing

The origins of engineering drawing can probably be seen in the plans produced for the large-scale building undertaken in early civilisations. We know that ground plans, drawn to scale, were in use by 2000 B.C. The Roman architect and engineer, Vitruvius, advocated the drawing of plans and elevations to instruct the craftsmen.

It is not supposed that the majority of craftsmen during the Middle Ages worked from drawings. The dimensions and shapes of the small variety of articles they were called upon to manufacture would be well known within the craft, and the guild secrecy of the time would be against permanent records which might be stolen. However, craftsmen such as the carpenter and stonemason were obliged to use drawing techniques to mark out their materials with instruments such as dividers, squares, and straight-edges.

The shipbuilder was also an exception in that he was required to produce drawings in order that the rib shapes could be marked out full-size on the lofting floor. The rib curvatures were composed of a series of circular arcs; the centres of the arcs being indicated on the drawing.

Technical illustrations appear in the writings of Leonardo da Vinci who used exploded pictorial views to illustrate some of his inventions. Agricola, in his *De Re Metallica,* incorporated many fine illustrations of mechanical appliances. Although all the drawings are pictorial he made use of sections, ghosted views, and detail drawings to explain clearly the constructions of the mechanisms.

The earliest example of the use of three-view orthographic projection is the work of Albrecht Dürer who, towards the end of the sixteenth century, used this method to indicate the proportions of the human figure in a textbook for artists. Dürer was also interested in perspective projection, making use of a viewing plane and eyepiece to compose some of his drawings.

A firm theoretical basis for engineering drawing was provided by the

Frenchman, Gaspard Monge, who published his book on descriptive geometry in 1795. Monge started his career as a fortifications draughtsman. He made the important fundamental deduction that a point in space could be defined by reference to two imaginary planes, one horizontal and the other vertical. From this he developed his system of coordinate geometry, parts of which we use today for projecting curves of intersection and solving problems concerning the oblique plane. Although Mongean geometry served as a basis for engineering drawing on the continent it made little impact in this country.

A system of orthographic projection which made use of elevational and plan views evolved in Britain during the industrial revolution, but there were no generally accepted conventions. Throughout the nineteenth century the engineering drawing was still predominantly a work of art, generous use being made of colour to identify the several components. Usually only the overall dimensions were placed on the drawing, tolerances and fits being left to the discretion of the craftsman. The old guild atmosphere of secrecy was still prevalent. In the late 1890s, when F. W. Lanchester became works manager of an engineering company in Birmingham, he found that important dimensions were omitted from drawings of the company's products. These dimensions had only been known to Lanchester's predecessor who had committed them to memory.

In Britain and on the continent the method of placing projected views became standardised by the mid-nineteenth century. We know the method today as first angle projection, and its use follows naturally on the method used by Monge for projecting points onto planes.

In America, isolated from European tradition, engineering drawing evolved from practical requirements rather than theoretical principles. It seemed logical to the American engineers that a view from the top of the objects should be placed above the elevation. Unrestricted by any conventions of projection they were completely free to use this method.

The modern methods of teaching the relationships between first and third angle projection, using the four quadrants and transparent planes, only appeared after both systems were well established. These methods were considered necessary in order to explain third angle projection in the terms of a rational system of projection.

Although some Americans urged the adoption of the European system for the sake of uniformity, they were unsuccessful. Third angle projection, as it is now known, was adopted throughout the United States and Canada – a decision which still creates problems today.

Additional Reading

P. J. BOOKER, *A History of Engineering Drawing,* Chatto & Windus 1963.

H. A. FREEBURY, *A History of Mathematics for Secondary Schools,* Cassell 1960.

Chapter 8
Craftsmen and Engineers

Origins of a Profession

The pyramids of Egypt and the Ziggurat temple towers of Sumeria provide ample evidence of engineering skills amongst the early civilisations. It is with the Romans that a really systematic approach to the whole subject is discernible. It might have been the military character of Roman life which was responsible for the high degree of organisation of Roman workers.

Closely knit corporations, or *collegia,* controlled each trade, and through them the skill of the *magistri* or craftsmen was passed on to the apprentices. Within the Empire there were many categories of employment which involved metallurgy and engineering. One of the most prized skills was that of the armourers, the *fabricenses,* who were branded to prevent them from changing their occupation. Roman coins were minted by the *monetarii,* and there were miners in Roman Britain working lead in Derbyshire, copper and gold in Wales, and tin in Cornwall.

There is no doubt that engineers of the Roman Empire often enjoyed a high status. Vitruvius, the inventor of the vertical water-wheel, stated that architects should be well born and soundly trained, the term architect embracing engineers. From the war dispatches of Caesar it is possible to gain an insight into the world of the Roman military engineering corps. Under the control of the Praefectus Fabrum they worked in large gangs, building siege towers and great war engines like the *ballista fulminalis,* which could send missiles hurtling over a river as wide as the Danube.

The Middle Ages

Much of the engineering knowledge necessary for the castle and church building of medieval times had descended from the civilisations of Arabia and China. It was the convention of the times to remain anonymous, but despite this we know the names of many of the great architect-builders, such as Ailnoth, who in the twelfth century worked on the Tower of London. He could work, so we are told, with equal facility in wood or lead, glass or stone. There was also James of St George and Richard Lenginour who built for Edward I of England the castles of Conway, Flint, Rhuddlan, and Chester to consolidate Edward's conquest of the Welsh.

Although these men would be described today as civil engineers, it should be remembered that our rather strict lines of demarcation were unnecessary until well into the nineteenth century. Much of the experience gained at this time was later to be of value to the mechanical engineer. The medieval builder often made a geometrical drawing of his assignment, and there are records even of a tracing house or drawing office on sites. Scale models were not used at this date; templates made of cloth, wood, or lead were employed by the builder to indicate to the quarrymen the shape of stone required.

Guilds

Metalworking, in common with all other crafts, was administered by the guilds, which were in many ways similar to the collegia of Ancient Rome. During the early period of their existence in the twelfth century, guilds were free associations of craftsmen who followed a common calling. The principles of sound craftsmanship, and protection against inferior commodities and cheap labour were carried out for the good of the community. The craftsman pursued his calling openly, in the sight of the people, a practice which persists still in oriental countries. To prevent competition it later became obligatory to join the guild, a practice which in today's trade union terms would be described as a 'closed shop'. By the end of the fifteenth century members were first required to satisfy a number of strict conditions.

Entry to a craft guild was usually by means of a seven-year apprenticeship. A master was not permitted to engage more apprentices than he could effectively train, and in a written contract or indenture he formally stated his obligation to give tuition to the apprentice in the appointed craft or 'mystery'. In addition he undertook to adminster to the apprentices spiritual and moral welfare, and the apprentice, in return, agreed to practice obedience, self-control, and fidelity.

To be accepted as an apprentice it was sometimes necessary to pay a substantial sum of money to the master. Boys commenced their apprenticeship around the age of fourteen, and many would be expected to sleep in the workshop. In London groups of apprentices often ran wild, and in an age without policemen constituted a social menace. In an effort to control them, an order by the Mayor of London in 1479 forbade 'all labourers, servants, and apprentices of any artificer (craftsman) or victualler to play tennis or to use cards or dice on pain of six days' imprisonment'. In 1517 some London apprentices were involved in a slight commotion, and defied the instructions of the city fathers who had ordered everyone off the streets after 7 p.m. Royal authority replied with ruthless reprisals, and a charge of treason was brought against the boys. 'Thirteen poor younglings', as the chronicler called them, were executed, 'to the horror of their parents and fellow apprentices'.

GUILD REGULATIONS

After seven years' training a young man would become a journeyman, this being the intermediate stage between an apprentice and a master craftsman. The expense of becoming a master craftsman, that is, one who owned his workshop, prevented many journeymen from achieving this status. The unsuccessful journeymen were later to form dissident groups within the guilds. Those who became master craftsmen were still subject to strict regulations aimed at preventing what was then considered unfair competition. To prevent members from making extra profit by cheating on quality, guild searchers were appointed to uncover shoddy or inferior products and report them to the guild court. Advertising was forbidden as it was thought to be an unworthy way of gaining advantage over a fellow craftsman, and it was held to be sharp practice to use anything but the traditional tools and techniques or to work by artificial light.

As the guilds became increasingly self-centred, attempts were made to restrict entry to the sons and dependents of guild members. Disgruntled journeymen were attracted to towns where guild regulations did not apply, and where they would be able to open their own workshops.

The Tudor governments recognised the importance of the guilds to a disciplined society. In an unsuccessful attempt to preserve the decaying guild system, the Statute of Artificers of 1563 stipulated uniform regulations for apprentices throughout the realm, which included a seven-year apprenticeship. Legislation could not prevent the decline of a system which had no place in a changing economic structure. The future belonged to the capitalist employer, or entrepreneur, who provided raw materials and sold the finished product, paying the craftsman for his labour.

Mine Engineering in the Sixteenth Century

By 1525 it is estimated that in Germany 100,000 men were concerned in the mining and processing of metal ores. The engineering aspect of this work was considerable, and the versatility of the mining engineer was fully described in a written work entitled *De Re Metallica,* published in 1556. The author was a German doctor whose pen name was Agricola. The work ran to twelve volumes, and in the sixth Agricola illustrated a large number of contemporary mechanical devices, including drawings of bellows, chain and piston pumps, and water-wheels. Descriptions are given of horse-driven whims or hoists, flywheels, and the use of wind power to operate ventilation fans.

German mining engineers came to Britain, and under the patronage of the Crown a number of companies were formed to exploit the country's resources of copper, zinc, lead, and silver. These schemes provided a partial solution to the acute financial distress which embarrassed the Tudor monarchs. State support was invaluable to an industry which was to amass a fund of technical knowledge from which engineering was to benefit in the future. There was also a direct connection between mining and the subsequent development of steam power. Drainage accounted for some three-quarters of the running costs in mining, and it was to solve this problem that the early steam engine was invented.

The Millwright

Water-mills and windmills constituted fairly complex mechanical structures. Constructed almost entirely of wood, they were erected and maintained by the millwright, an artisan who, although basically a carpenter, was the possessor of many other skills. He would have the knowledge and ability to undertake any repair work required on his journeys around the mills of the countryside. The work undertaken might range from the cutting of a millstone to the repair of gearing, and could well include smithying and turnery.

He would have a sound empirical knowledge of machinery and would be able to make calculations involving power and velocities. His skills extended to the building of bridges, constructing waterways, and cutting canals. By the nature of his work he was necessarily resourceful, and his varied skills would tend to place him outside the confines of a particular guild.

Sir William Fairbairn, the inventor of the Lancashire boiler, who started his career as a millwright, suggested that within this calling was centred most of the mechanical knowledge that existed at the end of the eighteenth century. The esteem in which the millwright was held in the early nineteenth century is reflected in the fact that he could earn the comparatively high wage of two guineas a week.

Engineering in the Factory

For many years after the invention of the steam engine it was the practice for engines to be built on the site. Usually only the cylinder, piston, and valves, were supplied by the manufacturers, and the supervision of the assembly and local construction of the other necessary parts, including the beam, would be entrusted to the millwright. By the end of the eighteenth century, as the engines had become more complex, the complete construction of the engine was undertaken in the factory.

Before work started on the production of a new engine in the Soho workshop of Boulton and Watt, the processes involved were carefully planned. Lists were compiled itemising the work to be carried out on the various components. The work then passed into the hands of the pattern-makers and moulders who were responsible for the foundry work, the blacksmiths who were responsible for all forge work, and two new classes of skilled workmen—the fitters and the turners. The fitter was concerned with the filing and scraping of castings, the mating of working surfaces, and the assembly of components. The turner was responsible for components which had to be produced on the lathe, and at a time when the lathe was still a simple machine, he had to be extremely skilful.

The Engineering Unions

The Combination Acts, which prevented the formation of Trade Unions during the Napoleonic Wars, had been repealed in 1824. It was natural that the new classes of skilled workers within the factories should seek wages comparable to those of the millwright whom they were gradually replacing. Initially, the factory system worked to their advantage, as they were in a position to combine in protection of their interests. Unions were formed representing particular trades and skills, such as machinists, boilermakers, and metal-planers. Many of them, like the Steam Engine Makers of Liverpool, were organised on a local basis. The first nationwide union was that of the Journeymen Steam Engine Makers, founded in 1826.

In the early days of the J.S.E.M. branch meetings were held in public houses, and there was a subscription of sixpence per week. The Union policy was to protect the craft status of its members and to try to improve wages and conditions. The employers on the other hand attempted to cut wages and lengthen working hours.

Some of the tactics employed by the unions would now be described as restrictive practices. An example of this was their attempt to control the intake of skilled men into any particular factory, so that the law of supply and demand would operate in their favour. By 1845 the unions were opposing strongly other practices which they considered unjust; the most important of these was the piecemaster system by which the employer contracted a price for a piece of work with the chargehand. Parts of the job would then be subcontracted to other workmen at a rate of payment decided by the chargehand. Exception was also taken to the employment of operatives on machines in place of skilled men, although some machines were simple to operate.

THE AMALGAMATED SOCIETY OF ENGINEERS

So long as the engineering workers were divided into a number of small unions their bargaining position was weak, and a growing body of opinion favoured the formation of one large engineering union. This occurred in 1851 with the formation of the Amalgamated Society of Engineers, Machinists, Smiths, Millwrights, and Patternmakers.

The A.S.E. differed radically from the earlier unions, the subscription of one shilling a week was comparatively high, but membership carried with it the advantages of a sick club, a funeral club, and a pension fund—welcome benefits in an age before social security.

The A.S.E., like its predecessor the J.S.E.M., adopted a militant and uncompromising attitude on matters affecting the status of its members. Almost immediately after its foundation it was involved in a bitter and prolonged fight with the Central Association of Employers of Operative Engineers. After one large firm had conceded the union demands for the manning of machines by skilled men only, an attempt was made to extract similar terms from the other companies. The masters retaliated with a 'lockout', and employees were not allowed back to work until they had signed a document renouncing membership of the union. Even in the mid nineteenth century the law of the land could not support an enforcement of this nature, and within a short time the union was enjoying an even greater measure of support. Despite the early stormy passage, the A.S.E. became a respectable organisation. Its members were skilled artisans, the sort of men who were likely to qualify for the vote under the Reform Act of 1867.

The industrial climate of the next forty years favoured the engineer. Threatened with 'go slow' tactics, employers were prepared to negotiate improvements, and the A.S.E. seldom had recourse to strike action. Wages increased in some cases by as much as 15 per cent, nevertheless there was a strike in Newcastle-upon-Tyne to demand shorter working hours. The employers tried to intimidate the men by bringing in German workers, but the immigrants were so frightened by colourful rumours about British working conditions that most of them returned to Germany. In January 1872 the nine-hour day was conceded, first in Newcastle and then throughout Britain.

Towards the end of the century the A.S.E. was the largest union in the country, with a membership of 90,000 and with ample financial reserves, but the engineering industry was changing to the detriment of the skilled man. The solidarity of the A.S.E. was threatened by the formation of rival unions representing the brass finishers, toolmakers, and patternmakers. With machinery becoming simple to operate, the newly formed Engineering Employers Federation was prepared to take a tough line. The employers were particularly determined to force the union to accept the manning of certain machines by unskilled workers. Friction between employers and workers led to a number of disputes which were firmly answered with 'lockout' notices. The situation escalated as the union retaliated by ordering a strike at all factories where a lockout policy was in operation. By the end of 1897 something like 50,000 men were unemployed. After a general stoppage, which lasted into 1898, the union funds were exhausted and the men were ready to return. They had to accept new conditions of employment which gave the owners much wider discretion over the fixing of pay rates. In addition it was laid down that the union should have no say in management and that engineering employers could direct a man to any job they chose—a dire blow to the union's claim to put a skilled man on every machine. Such terms indicate that the status of the craftsman in engineering had sunk quite low by the end of the nineteenth century. The workshops employed an increasing number of semi-skilled and unskilled workers, and to the disgust of craftsmen the lines of demarcation between one skill and another were progressively ignored by employers. This experience was not confined to Britain. In Germany a similar position arose where support for the skilled Hirsch-Duncker unions in the metal and engineering industries fell away rapidly during the 1890s.

The Professional Engineer

The progress of science in the seventeenth century owed much to bodies like the Royal Society in Britain, which was founded with the active support of Charles II, in 1660. Savery demonstrated his steam pump before the Royal Society in 1699, and in 1711 its French counterpart, the Académie Royale des Sciences, commissioned one of its members to examine every process of manufacturing then in existence. Also important was the fact that such societies encouraged the formation of bodies in the provinces, to promote not only science but manufactures and technical matters generally.

One of the most fascinating local groups was the Lunar Society of Birmingham, to which Watt, Boulton, Smeaton, and Joseph Priestley, the famous chemist, belonged. It met monthly at the full moon, and topics of discussion ranged from abstruse philosophy to the practical details of new processes which were often tested at the Soho works. Other groups, such as the Society of Millwrights which was founded in 1803, are clearly the ancestors of modern professional bodies.

From the very beginning of the Industrial Revolution, Britain and other developing countries needed men who could give to engineering the dignity of an accepted profession. This was not an easy task, for there were very definite views at that time on what were suitable occupations for gentlemen. Unfortunately, work of a practical nature was not usually considered to be one of them. The prejudice against which an engineer often had to contend is captured in the following quotation from *The Dialogues* by Max von Weber, the German author and engineer.

Count C. Dear Baron—I congratulate you on your son. He is quite a charming young man. I was astonished at the good knowledge of literature, art, and science, that he betrayed in conversation with fine tact and without pretension. Beautifully brought up. What do you proposed to make of him?

Baron E. He will be a technologist Count C. and will soon enter the Commercial Academy at B.

Count C. You are joking! With your ancient name, your connections in the best circles! This elegant young man, cut out for a diplomatic career, to be a sort of superior workman!—Forgive me if I laugh.

Despite this sort of attitude, which was also prevalent in England, ability could always command respect. James Watt, for example, was offered the position of Sheriff of the County of Warwick, and Henry Maudslay, a man of humble birth, could claim a duke among his many influential friends.

THE INSTITUTION OF CIVIL ENGINEERS

The most distinguished group of engineers in Britain at the beginning of the nineteenth century were the civil engineers, who had formed their own institution in 1818. The first president was Thomas Telford, who prompted the reading and discussion of learned papers within the society. Ten years later in 1823 the Institution of Civil Engineers received a royal charter.

Mechanical engineers, having no society of their own, were allowed to join, but the Institution was dominated by a fractious group of canal engineers. These men resented the more mechanically minded members who were employed on the railways. Quite correctly they assumed that such men would bring about the decay of their livelihoods on the canals.

THE INSTITUTION OF MECHANICAL ENGINEERS

In 1846 a group of railway engineers were holding engine trials on the Lickey incline near Birmingham. During a rainstorm they took shelter in a small hut and the conversation turned to an incident involving George Stephenson, the most distinguished of the railway engineers. The Institution of Civil Engineers had refused him membership unless he could give evidence of his engineering ability by means of a written essay. In view of Stephenson's achievements this was indeed a snub. They decided to form their own society, and from this informal gathering the Institution of Mechanical Engineers owes its origin. In the words of its Charter it was formed with the aim of advancing knowledge 'by a mutual exchange of ideas respecting improvements in the various branches of mechanical science'.

Not only were the railway engineers the founders of this institute, but they advanced the prestige of mechanical engineering through the work of such outstanding members as I. K. Brunel and Robert Stephenson, who later became public figures of great importance.

OTHER PROFESSIONAL INSTITUTIONS

The trend in the nineteenth century was towards a greater specialisation in all fields of engineering. After the reform of Parliament in 1832, for instance, there was considerably more interest taken in sanitation. It has been said that the early Victorian town had to be 'seen and smelled to be appreciated'. To deal with recurrent epidemics a new form of sanitary engineering emerged. The aim was to supply piped water to every home so that waste matter could be flushed away into deep sewers which would carry it to treatment plants.

Concern over sanitation was one of the main factors leading to the creation of the Institution of Municipal Engineers in 1873. Other specialist engineering bodies were also springing up at that time, including the Institute of Electrical Engineers, founded in 1871, the Marine Engineers in 1889, and the Heating and Ventilating Engineers in 1897. In the British cycle industry an enterprising group of engineers formed their own association which later expanded to cater for engineers in the automobile industry.

Engineering Training in the Nineteenth Century

It is remarkable that Britain produced so much talent and ingenuity during the Industrial Revolution without any coherent national policy for training craftsmen or engineers. Apprenticeship, which guaranteed at least a rudimentary tuition, fell into disrepute during the eighteenth century. The new manufacturing industries such as cotton were not covered by guild regulations. Employers had a growing dislike of restrictions, and these included the rules and regulations which governed apprenticeship. In 1814 the main clauses of the Statute of Artificers, which had imposed a national system of apprenticeship, were repealed. Although the law had been widely ignored, its very existence had indicated a favourable attitude towards apprenticeship by the government. It was now legally permissible for a man to practice any craft without an apprenticeship.

Nevertheless apprenticeship was continued, and occasionally, as in the firm of Boulton and Watt, it was of a high calibre. The majority of firms, however, offered little in the way of training for apprentices, and in some industries an alternative to apprenticeship was adopted. In steel mills and iron mills a system known as 'progression' became the practice. Starting with light manual work a boy would graduate, over the years, up a prescribed ladder of promotion until he became a first hand on the furnace, or a roller-man, positions which were well paid and combined skill with seniority.

APPRENTICESHIP IN OTHER COUNTRIES

Unlike Britain, where government policy was against interference in private industry, the German state of Prussia took a more enlightened attitude towards apprenticeship, which was seen as a method of stimulating industrial activity. State decrees established a system of apprenticeship, in small and large firms alike, which demanded a high standard of training. The numbers of apprentices were controlled, and special trade schools were founded to operate in conjunction with technical high schools and to supply the type of secondary education necessary to future engineers and craftsmen.

At the beginning of the present century the U.S.A., and in particular the state of Wisconsin, followed the German lead, and obligatory minimum standards for apprenticeship training were clearly stipulated.

HIGHER TECHNICAL EDUCATION

In the early years of the nineteenth century a Cambridge professor, the Rev. William Farish, instituted a series of lectures which included the construction of machinery and the study of metals. The lectures were illustrated by the use of models to demonstrate primary mechanical functions. It was to simplify the erection of the models by his students and assistants that Farish developed the use of isometric projection. In 1813 he was made Professor of Natural Philosophy at Cambridge. Among the eminent successors to this chair was Robert Willis, famous for his work on gearing.

Apart from this notable exception at Cambridge, courses in professional training were slow in developing. Throughout the nineteenth century the majority of British professional engineers received their training as fee-paying pupils of industrial firms or consultant engineers.

It was considered essential that the pupil engineer should be practised in the use of hand and machine tools; professional knowledge was acquired by working under a qualified engineer. The few engineers who received their training at university or college were viewed with suspicion in the industry.

During the French Revolution, a group of prominent scientists, including Lavoisier, Monge, and Carnot, was concerned to reduce the lead that Britain had taken in industry. To this end the Ecole Normale was established in 1794 as a training ground for teachers. Entry was by ability rather than birth, and the syllabus included practical and theoretical training in the engineering sciences. The method of instruction was based on the descriptive geometry devised by Gaspard Monge who was a professor at the Ecole. Under its later name of the Ecole Polytechnique the school established a high reputation for the quality of its engineering training.

In America the famous Massachusetts Institute of Technology was founded at the end of the Civil War, while in 1862 funds were made available for the land-grant colleges where the accent was on mechanical and agricultural education.

The promotion of higher engineering studies by the direct encouragement of the state can best be seen in the case of Prussia. There was no one in Britain to compare with Peter Beuth (1791-1863) who led the Prussian Government Technical Commission in 1819. It was by his influence that lectures on technology became popular, and he encouraged the publication of technical books, which led to the establishment of well-known German technical publishers such as Springers of Berlin. Beuth's industrial institute in Charlottenburg later expanded to become a famous technical university.

While Britain had nothing to compare with the state-directed technical education of Prussia, an attempt was made to provide some form of technical education for the workman. As early as 1800 in Edinburgh, George Birkbeck, a medical practitioner, was running a series of lectures devised to appeal 'to those engaged in the practical exercise of the mechanical arts'. Others followed the lead until by 1841 an estimated membership of 120,000 attended some 700 Mechanics Institutes throughout Britain. Unfortunately, in a country without a system of primary education on which to build, these efforts met with only a limited success.

The prospect, however, of expanding local technical institutes of this type appealed to the more progressive authorities in the larger cities, and they were supported by farsighted industrialists, who were both proud of their cities and acutely aware of the need for more technical education. In 1851 a benefaction left by a nonconformist minister, John Owens, was used to found a technical college in Manchester, later to become the first civic university and known as the Victoria University. It was closely followed by the establishment of universities in other industrial centres such as Leeds, Liverpool, and Birmingham.

No less important was the move to obtain the acceptance of engineering as a reputable university subject. William Ritchie, Professor of Physics at the University of London, successfully appealed to the University Senate in 1832 to allow him to give additional lectures in engineering. Glasgow University appointed a Professor of Mechanical Engineering in 1840. In 1851 this chair was occupied by Professor Rankine, whose textbooks on steam engines and applied mechanics became standard works on these subjects. In 1856 a Professor of Mechanical Engineering was appointed to King's College, London, and in 1867 Whitworth helped to establish the first chair of engineering at Owen's College. In addition Whitworth endowed a number of scholarships for aspiring engineers, a lead which, unhappily, was not copied by many others.

TECHNICAL EDUCATION AT THE SECONDARY STAGE

University engineering could be only partially effective when there was the void in our educational system which secondary schools fill today. An important step in the right direction was taken in 1872 when the Society of Arts instituted a system of examinations with which to test the first stages of technical and craft knowledge. There were, however, insufficient centres where instruction for these examinations could be given.

The government, concerned at the progress being made abroad, appointed a Royal Commission in 1882 to investigate the state of technical education in this country. After sitting for two years the Commission recommended the creation of secondary schools for the purpose of giving initial technical training. Soon after the Commission had reported, a new system of local government was approved by Parliament. This provided for county councils and county borough councils, democratically elected, which were ideally placed for implementing the recommendations of the Royal Commission.

Additional Reading

W. H. G. ARMYTAGE, *A Social History of Engineering*, Faber 1961.

G. D. H. COLE and RAYMOND POSTGATE, *The Common People*, University Paperbacks 1968.

F. KLEMM, *A History of Western Technology*, Allen & Unwin 1959.

L. T. C. ROLT, *'The Mechanicals': Progress of a Profession*, Heinemann 1967.

Biographical Sketches

Iron and Steel

The Darbys of Coalbrookdale
John Wilkinson
Sir Henry Bessemer
Sir William Siemens
Sidney Gilchrist Thomas

Steam Power

Thomas Newcomen
James Watt
Richard Trevithick
The Hon. C. A. Parsons

Machine Tools

Joseph Bramah
Henry Maudslay
Sir Joseph Whitworth
James Nasmyth
Brown and Sharpe

Transport

George Stephenson
Robert Stephenson
I. K. Brunel
Henry Ford

The Darbys of Coalbrookdale

The success of the early Industrial Revolution depended largely on cheap and plentiful supplies of iron smelted with coke. Abraham Darby I (1678-1717) was the pioneer who made this possible, and who founded at Coalbrookdale in Shropshire a famous ironworks which became a centre of industrial development.

When Darby was born in 1678 at Dudley in the English Midlands, iron-making was already well established there, and most of the townspeople were aware of the abortive efforts of a local worthy, Dud Dudley, to smelt iron with coal instead of the conventional charcoal. The young Darby was fortunate to be born into a Quaker family, for the Quakers were well known for their enterprise and their wide connections in the business world. He was apprenticed to Jonathan Freeth, a maker of malt mills in Birmingham. We know that at this time maltsters were beginning to convert coal into coke for use as a fuel, and Darby undoubtedly took note of this practice during his apprenticeship. In 1699 he set up in partnership with other Quakers in Bristol and became interested in the manufacture of brass cooking pots. Darby was of the opinion that these pots—the sort which hung over a fire—could be more effectively made of iron, and this idea led him to the important technical innovation of casting in sand rather than in loam. He obtained a patent, which was granted for a period of fourteen years from April 1707, and the patent document catalogued all the advantages 'to the poore of this owne kingdom' from a technique 'by which iron pots and other wares may be cast fine and with more ease and expedition'.

Casting iron pots in sand foreshadowed the wider patent for sand casting taken out by Isaac Wilkinson in 1758 and heralded an era of foundrywork upon a much larger scale. More immediately it led Abraham Darby into certain differences of opinion with his colleagues and made him determined to specialise as an ironmaster. With this end in view he went to some trouble to find a good site, and in 1709 moved to Coalbrookdale. It is almost certain that Darby came to Coalbrookdale with the express purpose of smelting iron with coke because we know that the earliest experiments took place in 1709, very soon after his arrival. His daughter-in-law later wrote 'He first tryed with raw coal as it came out of the Mines but it did not answer. He not discouraged, had the coal coked into Cynder, as is done for drying Malt and then succeeded to his satisfaction.'

To effect his process, Darby was able to adapt an existing blast furnace which was about 6 m high. He probably needed a stronger blast of air for there is a record of his purchasing a new pair of bellow boards.

The output of Coalbrookdale in these early years amounted to something between five and ten tonnes a week. Much of this was cast into cooking pots or into pigs of iron. Darby sold the cooking pots at fairs, and the pigs were sold direct to small foundries. This meant that the middlemen, who had previously collected a disproportionate share of the profits were cut right out. By the time of his death in 1717 Darby's shrewdness had caused his business to prosper, but the use of coke for smelting iron did not become widespread until many years afterwards.

Abraham Darby was the founder of a dynasty of ironmasters who were to make Coalbrookdale famous. He was succeeded for twenty-eight years by his son-in-law, Richard Ford, who increased furnace capacity and obtained many orders to provide pipework and cylinders for Newcomen engines.

Under Abraham Darby II, who managed the works from 1739 to his death in 1763, coke was accepted for the first time in the production of bar iron. The contribution of the Darby family was once again extremely significant, for this commodity became basic to the rapidly expanding nail trade of the Black Country, as well as to a variety of other manufactures.

After the death of Abraham Darby II the management passed to his son-in-law, Richard Reynolds. Reynolds replaced the wooden track used for internal works transport with iron rails. This gave Coalbrookdale the first permanent way ever laid.

Even at this mature stage in its history Coalbrookdale continued to be closely concerned with advanced technical developments. In 1779 Abraham Darby III, who had become head of the firm eleven years previously, cast the members of the world's first iron bridge. It was erected over the river Severn at Coalbrookdale in the year 1779 by the Shrewsbury architect, Thomas Farnolls Pritchard, and it stands today, as significant a national monument as Stonehenge or Westminster Abbey.

As Quakers, the Darbys voluntarily abstained from the profitable activity of cannon-making, a sacrifice which obliged them to modernise wherever possible in order to remain competitive. For this reason they once again anticipated other ironmasters in 1802 by ordering a high-pressure steam engine, designed and installed by the famous Trevithick.

But the continuity of one Quaker family at Coalbrookdale meant more than purely technical progress. The Darbys built houses and schools, and interested themselves in the welfare of the local people. Their kindness evoked respect and affection, and this was clearly demonstrated at the celebrations which were held to mark the birth of Edward William Darby in 1850. 'Hail, child of our hopes, many happy years be thine' pronounced the banners of the local people, 'Long live the name of Darby'.

John Wilkinson, 1728-1808

'But before I proceed any more with my lingo
You shall all drink a toast in a bumper of slingo;
Fill up, and without any further parade,
John Wilkinson, boys, that supporter of trade.'

It was with these words that John Wilkinson's workmen sang the praises of their illustrious master: a pioneer in the use of iron, an inventor and tool-maker of originality, and one of the greatest industrialists of his time. Wilkinson was born in 1728, the son of Isaac Wilkinson, an ironmaster who had established his own foundry at Lindale in Lancashire, and then moved to take over control of the Bersham foundry at Wrexham.

John and his brother, William, played a considerable part in developing a new method for casting cannon in sand, using iron moulding boxes, originally patented by their father in 1758. Sand casting had been invented by Abraham Darby I, but it was only with the Wilkinsons that it became the standard method for dealing with large and tricky projects like cannon, cylinders, pipes, and sugar rolls. In the words of the Wilkinson patent, these could now be cast 'in a much more Neat, Compact, Exact, and Useful as well as Cheap and Expeditious manner than any Method hitherto known and made use of'.

John Wilkinson's rise to fame as an ironmaster was due, in part, to his readiness to use coke for smelting iron. He left his father in 1755 to work first in Wolverhampton, and then in the employ of a Mr Hoo of Bradley, near Bilston in Staffordshire. He married twice, and it is likely that on both occasions he was influenced in his choice by the capital his spouse could put into the business. Through a mixture of industry and ruthlessness he was, by 1763, complete master of a foundry at Broseley near Coalbrookdale, and by 1770 he was also in possession of the foundry at Bradley and had important business interests in North Wales at Bersham and Brymbo.

At Bradley, Wilkinson specialised in the production of cannon for which his firm gained a wide reputation. It was this interest in ordnance that led him to his famous patent of 1775 for the accurate boring of cannon (see chapter 4). This important invention anticipated the era of machine tool building which was to begin with Maudslay.

Its main application was in boring the cylinder for the Watt steam engine at a time when these engines were ready to make an entry into a receptive market. Wilkinson cast and bored the cylinders to the specifications of Boulton and Watt at Soho, and then transported them directly to the customer. The understanding between the partners and Wilkinson deteriorated, and perhaps because of the latter's imperious nature, Boulton and Watt decided to establish their own foundry at Smethwick in Staffordshire. Nevertheless, Boulton, a judicious businessman and a good judge of character, had a firm respect for the great ironmaster. 'I can't say but that I admire John Wilkinson', he wrote, 'for his decisive clear and distinct character, which I think is first-rate of its kind.'

Another of Wilkinson's interests was in lead mining, and this led him to promote a scheme for the manufacture of lead pipes. The production technique involved the use of a drawbench on which pipes were drawn through a series of dies. In 1762 Wilkinson arrived at an agreement with the Coalbrookdale company, who made a great deal of industrial pipework, to fix prices. The agreement also provided for uniform sizes of pipe, which suggests that on the question of standardisation as well as on other matters, Wilkinson was way ahead of his time.

In 1792 he introduced a steam-driven rolling mill, a most important innovation, especially in the light of later developments, which increased the demand for rolled products. The Wilkinson mill weighed about eight tonnes, and power was provided to the top roll only.

Before the use of coke in iron smelting, blast furnace and foundry needed to be closely connected, and both had to be near a source of charcoal. Coke smelting brought with it an independent status for the foundry as long as it could be supplied with fuel, and John Wilkinson improved the flexibility of the iron industry even further with his furnace of 1794. The furnace, 3 m high and clad with iron, was designed for converting pig iron. It was the prototype from which William, John's brother, developed a general foundry cupola which could melt down pig and scrap iron for casting into the form required.

As an advocate of the use of iron, Wilkinson had no equal. He built an iron chapel for the Methodists near his works at Bradley, and at his home overlooking Morecambe bay, he had an iron coffin ready for his own occupation. Such eccentricities were not without their powers of self-advertisement, but the real turning point in Wilkinson's campaign to promote his product came in 1787 when he attempted to float *The Trial,* the first iron vessel in history. The barge weighed eight tonnes, and the doubting Thomases turned out in droves to see the impending fiasco. They were disappointed, and Wilkinson later wrote: 'Yesterday week, my iron boat was launched. It exceeds all my expectations, and convinced the unbelievers who were nine hundred and ninety-nine in a thousand.'

William Wilkinson played a great part in the establishment of the iron industry in France and Prussia, but it was John Wilkinson himself who supplied the 64 km of iron pipes which were used by the French engineer, Perier, to bring water to Paris.

The two brothers quarrelled as young men, for there were few people whom John Wilkinson did not succeed in annoying at one time or another. Yet although short-tempered, he was an outstanding figure in a period noted for its fascinating characters. He fathered three children after the age of eighty, and so great was the magnetism of his personality that the people of Bradley were convinced that he would reappear seven years after the day of his death, mounted on a grey horse.

Sir Henry Bessemer, 1813-98

The name of Henry Bessemer became world-famous through the merits of one great invention—the steelmaking converter—but this was only the crowning achievement of a truly remarkable and ingenious man. Bessemer was born in January 1813 at Charlton, Hertfordshire, into a sound middle-class household. His father did not encourage his interest in the family type-foundry business, but, faced with the boy's persistent curiosity and precocious practical talents, he relented, and Henry left school at fifteen to serve an apprenticeship.

Throughout his long life Bessemer was to turn his mind to many things. One of his early ideas was an embossing die which could stamp designs on cardboard. He also made valuable suggestions on improving stamp production, and on speeding up the printing of newspapers. Then, while still a young man, Bessemer invented a mechanical method of manufacturing 'gold' powder. This supplanted the traditional process, and as 'gold' powder sold at seven shillings an ounce the profits made Bessemer a rich man.

In 1853 he was on the threshold of his greatest discovery. Increasingly concerned over the inferior quality of British cannon being made for the Crimean War, he prepared designs which were ignored by shortsighted War Office officials. Bessemer then approached the French military authorities, who were interested in his proposals but had serious misgivings about the strength of the cast iron from which Bessemer's guns would be made. With these criticisms in mind the inventor returned to England, determined to make iron which would have the strength required for his guns.

The aim, as Bessemer himself expressed it, was to produce 'a superior sort of cast iron by removing the carbon and other impurities to give the characteristics of malleable iron'. To achieve this he built himself a version of the reverberatory furnace. The small tuyeres or air vents which were incorporated increased the heat of the furnace by using a blast of air. Accidentally some pigs became lodged against the tuyeres, and Bessemer found that the air from the tuyeres had completely decarburised them, at the same time causing an exothermal or heat-producing reaction.

This clue encouraged the inventor to construct a cylinder, 1·25 m in height, at his workshop in St Pancras, London. Air was introduced through six tuyeres at a pressure of between 0.7 and 1.05 kg/cm^2. After ten minutes of pyrotechnics which were to become a feature of the Bessemer process, the metal charge in the container quietened. When tapped off, it was found to be malleable and decarburised. It was necessary, however, to make one further step, and a short while after the invention of the converter a significant discovery was made by Robert Mushet. The latter showed that by adding a small quantity of ferro-manganese the surplus oxygen in the metal would be absorbed, and at the same time the final carbon content could be adjusted to give steel.

It had not been possible hitherto to produce steel on a tonnage basis, and when Bessemer announced his findings in a paper to the British Association for the Advancement of Science there was immediate interest amongst the ironmasters. In one month alone Bessemer made £27,000 by issuing licences for the use of his process. Disillusion, alas, soon followed when ironmasters on all sides reported complete failure in their results, and Bessemer was accused of being a fraud.

It took him two years' hard work to find out that the cause of the trouble lay in the phosphorus content of the pig iron. He himself had fortuitously used a low phosphorus pig iron from Sweden for refining, whereas the iron-founders had used the native high phosphorus iron ores. As a result of these unfortunate teething troubles, the Bessemer process was seriously under-estimated in Britain. Bessemer himself amply justified the process by producing steel at his own works in Sheffield at £15 a tonne cheaper than that made by the traditional method. But the inventor became rather like the biblical prophet, honoured everywhere but in his own country. Abroad, his converters were quickly accepted, especially after the invention of the basic lining by Gilchrist Thomas. In Britain Bessemer had to engage in a prolonged struggle with traditionalists in structural engineering, and with the Admiralty who were large users of wrought iron, and who refused for some time to accept that the new steel was strong enough for their purposes.

Eventually, belated recognition came in the form of a knighthood, but neither the fame nor the abuse he received interfered with Bessemer's will to carry on working. One of his discoveries patented in 1856 demonstrated once again the farseeing nature of his genius. The patent concerned the use of oxygen rather than air in the Bessemer process, and it is obvious that the inventor realised that oxygen was the active agent. He did not follow up this idea, but we should remember that here he anticipated the modern L/D process which was not perfected until after the Second World War.

Bessemer, who died in 1898, showed throughout his life not only inventive powers of the highest order, but a readiness to experiment in many other branches of technology. It is this versatility above all which makes his career so amazing when viewed from our own age of narrow specialisation.

Sir William Siemens, 1823-83

Born in Germany, the seventh of fourteen children, William Siemens typified the nineteenth-century ideal of self-help and the full use of personal talents. When his parents died William was sixteen years old, and his education was directed by his elder brother, Werner, who was to become the head of the world-famous electrical company of Siemens Halske. After reading for a diploma in engineering at Göttingen University, William was sent to England to sell a special process for electroplating invented by his brother. This was in 1842, and Siemens, only nineteen years old, was still undergoing his practical works training as an engineer.

Despite his youth and the fact that he spoke little English, Siemens drove a hard bargain with the Birmingham firm of Elkington, who bought the process for £1600. He returned home convinced that in Britain, the leading industrial nation of the age, lay the best opportunities for developing his talents.

His training completed, he left again for Britain where he hoped to sell other inventions conceived by his brother and himself. These included an improved steam engine governor. Siemens was not so fortunate this time and struggled for some years until, in 1852, he invented a water meter which was an immediate success. The water meter was a simple but effective device consisting of a drum which, when rotated by the flow of water, registered the consumption in gallons on a dial. At this time doctors were urging the installation of piped water supplies to prevent the frequent cholera epidemics in large towns. Siemens's invention made water supply commercially acceptable so that, as well as proving extremely lucrative, it was of significant social importance.

By now Siemens was also a partner and British representative of Siemens Halske, and his combined interests assured for him a position of relative affluence. He was able to give his full attention to the problem of heat regeneration which had always intrigued him. Regeneration was concerned with the utilisation of waste heat to reduce fuel costs. The principle of heat regeneration had been used by Dr Robert Stirling, a Kilmarnock clergyman, who incorporated a regenerator in the closed cycle hot-air engine which he patented in 1816.

Encouraged by his brother Friedrich, Siemens built a modified reverberatory furnace in which the incoming air was preheated by passing through checker chambers previously warmed by the hot waste gases. This furnace was used to heat wrought iron billets, and achieved a reduction in fuel costs of 70 to 80 per cent.

Friedrich obtained a patent for the process in 1856, and it was later applied successfully to glass furnaces. The process was improved by William's invention of the gas producer in 1861 which enabled coal gas made from cheap coal to be used as a fuel. The furnace was redesigned so that the incoming gas and air were preheated in separate checker chambers and united at the edge of the hearth where they produced a very hot flame. The heat produced by this flame was sufficient to reduce iron and steel to its fully molten state, a task which the puddling furnace did not accomplish.

In France, Pierre and Emile Martin used a Siemens regenerative furnace to produce steel of a controlled carbon content by melting a charge of pig iron and wrought iron scrap, a process which they patented in 1865.

William Siemens was also interested in using the furnace to produce steel, and used the furnace for the production of crucible steel with some success. When, however, he attempted to refine pig iron in the hearth he found that the furnace lining burned away long before the impurities (carbon, silicon, etc.) were removed. He solved this problem by adding iron ore to the molten pig iron so that the oxygen in the ore would hasten the reduction of the impurities, Eventually the Martin method of using scrap to reduce the time taken to oxidise the impurities was adopted. In recognition of the parts played by the different men we now refer to this method of making steel as the Siemens-Martin open-hearth furnace.

The open-hearth process for producing mild steel became available at a time when the Bessemer process was discredited in the eyes of British ironmasters. The new process was cheap to run and was not dissimilar to the familiar reverberatory furnace to which the traditionalist ironmasters were accustomed. For this and other reasons the furnace was adopted readily by the industry, and became the major steel-producing furnace in Britain.

In 1869 Siemens entered the steel industry himself, just as Bessemer had done. He opened a plant at Landore in Swansea which contained twenty-four open-hearth furnaces; parts of this historic steelworks were still in production until very recently.

William Siemens continued to play an active part in the electrical side of the family business, particularly in the field of telegraphic communications. An honoured citizen in the country of his adoption, he was knighted shortly before his death at the age of sixty-one.

Sidney Gilchrist Thomas, 1850-85

Initially, both the great steelmaking processes—Bessemer and open-hearth—could only operate with non-phosphoric iron ores. This was a major drawback, for the larger iron-ore fields in Britain could not be exploited. The production of steel was also far from being a British monopoly, and there were important deposits of phosphoric ore abroad, particularly the one in Alsace-Lorraine which had been annexed by Germany in the Franco-Prussian War of 1870.

This problem was to be solved through the determined efforts of a gifted amateur, Sidney Gilchrist Thomas, who was helped by his cousin, Percy Gilchrist, an ironworks chemist at the Blaenavon works in South Wales. Sidney Thomas had enjoyed a sound middle-class childhood and had been educated at Dulwich College in London. The boy's prospects seemed blighted when his father died, and at eighteen he was obliged to take up a junior clerkship in the Metropolitan Magistrates' Court at Stepney. His original intention had been to follow a scientific career, and this led him to begin evening classes at Birkbeck College. A deep religious conviction and a sense of social awareness made him keen to use his learning for philanthropic ends. Thus when a lecturer at Birkbeck glowingly portrayed the financial rewards which might follow from the elimination of phosphorus in the Bessemer process, Sidney Thomas saw that this was a way of raising money for the benefit of others.

Between the years 1871 and 1875 he laboured tirelessly for a theoretical solution to this difficult problem, in addition to his heavy programme at the police court. The addition of lime might have been expected to eliminate phosphorus; unfortunately, at high steel-making temperatures lime could not be used because it always made a direct attack on the silica or 'acid' lining of the converter. Sidney Thomas saw that the answer lay in the firebrick lining of the converter which he thought should be of a chemically basic material to resist the action of the lime. It was here that his cousin, Percy Gilchrist, was of invaluable help, for he tested this conclusion in a long series of practical experiments. It was decided that the basic lining should be of dolomite, calcined or burned to resist the high temperature of molten steel.

To convince the sceptical world of metallurgy Thomas needed to test his lining under practical conditions, and this he was able to do through the cooperation of Percy Gilchrist's supervisor, Mr Martin, who was manager of Blaenavon.

In March 1878 Thomas, his work fully covered by patents, was ready to let the world know of his discovery. He had hoped to read a paper on the topic at a convention of the Iron and Steel Institute, but rather suprisingly he was passed over. At the meeting, however, Thomas met E. W. Richards, manager of the Bocklow-Vaughan works at Middlesbrough. Richards was struck by the importance of the discovery, and with his help the first steel-making in a Thomas converter took place in April 1879.

Within a short time Thomas's invention was being eagerly adopted throughout the world. On the continent they also used the inventor's idea of making fertilizer from the phosphoric slag. In the United States alone the patent rights were sold for 250,000 dollars, and at last Thomas had the money for his philanthropic schemes. It was particularly appropriate that much of the money was used to remedy the evil effects suffered by people who worked with phosphorus in the match industry.

The years of evident overwork broke Thomas's health, and he died of tuberculosis in 1885 at the age of thirty-four. As he himself put it in a letter to a friend in 1878: 'We have solved the greatest industrial problem of England.'

Joseph Bramah, 1749-1814

In 1770 the young Joseph Bramah came to London from Barnsley in Yorkshire, where he had served an apprenticeship as a joiner. His natural inventive talent was first applied to the design of a water closet which, by an efficient system of valves and traps, prevented stench and leakage. It remained in use until the end of the nineteenth century and did much to eliminate disease caused by inadequate sanitation.

This first venture gave Bramah a lasting interest in hydrostatics, in which field he was to make a number of significant inventions during the course of his career. The most important of these was the hydraulic press, the principle of which had been explained by Denis Papin some 100 years previously.

In 1784 he patented a lock which was advertised as being burglar-proof, and a reward of 200 guineas was offered to the first person to pick it. The reward was not claimed until 1851—thirty-seven years after Bramah's death—when an American locksmith took over fifty hours to accomplish the task.

The functioning of the lock depended upon the accurate construction of its working parts. To ensure this accuracy Bramah decided to construct machines to replace most of the manual operations and it was to build the machine that Maudslay, a young blacksmith, was engaged. By so doing Bramah was to give one of the foremost figures in technological history the opportunity to develop his talents.

Little is known concerning the relationship between these two outstanding men. Maudslay's later eminence and the fact that he left Bramah's employ after being refused a rise in his salary of thirty shillings a week have tended to obscure the fact that when he went to Bramah he was only eighteen, and singularly lacking in experience.

Probably the most important invention to emerge from Bramah's workshop was the lathe slide rest, almost certainly a joint project between master and assistant. Although not of robust construction it was capable of holding the tool in a fixed position and also had vee-slides that were adjustable for wear.

In 1797 Bramah patented a method of extruding lead pipe which did not enjoy a great vogue during his lifetime, but after its modification by Thomas Burr of Shrewsbury it became an important industrial process.

Bramah engaged Joseph Clement as his chief draughtsman and superintendent in 1813. Clement who later worked under Maudslay, was an engineer of great originality and he played an important part in the development of the lathe, the planing machine and of taps and dies for screwcutting.

At a time when precision was becoming important in engineering Bramah set a new standard of craftsmanship. A man of wide inventive talent, he included among his many practical schemes a wood-planing machine, a beer pump, and an automatic device for numbering bank notes. Oddly enough, considering his technical ability, he was an implacable opponent of James Watt, and held that the separate condenser added nothing to the Newcomen engine.

Towards the end of his life hydraulics became his consuming passion. He spent much time advocating the installation of high-pressure water mains in the streets, to be used for fire-fighting and the operation of hydraulic machinery.

Henry Maudslay, 1771-1831

An outstanding figure in the history of technology, Maudslay began his career filling cartridges at the Woolwich Arsenal when twelve years old. After a period in the woodwork shop he obtained a transfer to the smithy where, by the age of eighteen, his skill had earned him recognition which extended well beyond Woolwich.

The inventor, Joseph Bramah, was at that time seeking a highly skilled craftsman to assist him in the design and construction of machines vital to the manufacture of his complex tumbler lock. Maudslay was recommended to him, but Bramah was perturbed by the blacksmith's extreme youth and feared the reaction of his older craftsmen to a novice who had not served a recognised apprenticeship. Bramah's doubts were removed, however, when Maudsley demonstrated his competence by completely renovating an old bench vice in the space of a few hours. This was the beginning of an association that lasted ten years, and provided Maudslay with the environment in which to develop his unique standards of craftsmanship.

The lathe slide rest was the most important of a number of technical innovations which resulted from the collaboration of these two men, a device which, in the words of L. T. C. Rolt 'converted a crude tool into a precision instrument'. Maudslay claimed the credit for the invention of the cup leather washer which turned Bramah's hydraulic press into an important technical achievement.

In 1797 Maudslay left Bramah's service to establish his own business at Wells Street in London. We are told that this came about after Bramah had brusquely refused to raise his wage of thirty shillings a week. Shortly after his arrival in Wells Street Maudslay was asked by Marc Brunel, father of the famous I. K. Brunel, to undertake the manufacture of block-making machinery for the Navy.

These machines, forty-four in all, were to mechanise the production of the annual requirement of 130,000 pulley blocks. Although Marc Brunel conceived the original designs, their practical interpretation was due to the skill exercised by Maudslay who spent eight years on the project.

While occupied on this contract, Maudslay developed his revolutionary screwcutting lathe which was the basis of the modern centre lathe. Maudslay's work in connection with the lathe is described in chapter 4, and must be considered as his most important contribution to production engineering. As a result of the block machinery contract Maudslay was able to expand his business and begin the long and fruitful partnership with Joshua Field, a former draughtsman at the Portsmouth dockyard.

In 1810 the firm, now called Maudslay and Field, moved to a spacious site at Lambeth Marsh, Pimlico, where it was to flourish under its later name of Maudslay, Sons and Field, as one of the leading engineering firms of the nineteenth century.

Nasmyth, who was for some time Maudslay's devoted assistant, gave an attractive picture of him in these years as the good-natured employer who was always willing to take off his coat and show how a job should be accomplished. 'To be permitted to stand by', wrote Nasmyth, 'and watch the systematic way in which Mr Maudslay would first mark or line out his work, and the masterly manner in which he would deal with his materials and cause them to assume the desired forms, was a treat beyond all expression. Every stroke of the hammer, chisel, or file, told as an effective step towards the intended result. It was a never-to-be-forgotten lesson in workmanship in the most exalted sense of the term.'

There was more than just vanity in these demonstrations, as some of the foremost engineers of the nineteenth century were trained by Maudslay, and were to catch his infectious enthusiasm. Famous amongst his pupils were Clement, Roberts, Whitworth, and Nasmyth, all of whom made notable contributions in the field of machine tools. Lesser known were Napier, the prominent armament manufacturer, and Seeward, a well-known manufacturer of canal and shipyard equipment.

Maudslay's interest in the production of accurate screwthreads prompted him to manufacture his own taps and dies. He designed a range of standard screwthreads for use in his workshops with thread pitches suited to the diameter of the screw. This work undoubtedly influenced Whitworth in his successful attempt to standardise screwthread practice. Whitworth's advocacy of true plane surfaces also reflected the importance Maudslay attached to this aspect of production engineering. Nasmyth tells us that the practice of preparing surface plates in threes and the removal of high spots by hand scraping also originated at Maudslay, Sons and Field's. The Whitworth measuring devices which were so important to precision engineering in the later nineteenth century no doubt had their origin in the large micrometer used by Maudslay.

Maudslay would hardly have approved of our age of narrow specialisation for he was absorbed wholeheartedly in a great range of interests which included calico printing, coin-making, gun-boring, bronze foundry equipment, and hydraulic presses. His Lambeth works became famous for the outstanding marine engines which he pioneered. His first compact table engine appeared in 1815, and he dispensed with the characteristic beam of the Watt engine, driving the crank directly from a crosshead. By 1829, two years before Maudslay's death, ships' engines developing 300 kw were being produced by the firm.

Henry Maudslay who started as a 'powder monkey' filling cartridges, lived to become one of the most prominent industrial figures of his time, numbering among his friends many of the most famous contemporary scientists and engineers. He died, in 1831, of pneumonia contracted when visiting a sick friend in Boulogne. Characteristically, he was buried in an iron tomb of his own design. Its epitaph described him as 'eminently distinguished as an engineer for mathematical accuracy and beauty of construction'—a fitting tribute to his greatness.

Sir Joseph Whitworth, 1803-87

Joseph Whitworth, who was to become a key figure in nineteenth century engineering, was born in Stockport, Cheshire, where his father ran a school. It was decided that the boy should go into business, and at the age of fourteen he entered his uncle's cotton mill in Derbyshire. The office routine bored Whitworth, and he spent as much time as possible studying the machinery in the mill. He quickly mastered the construction and working of all the machines, and it is interesting to note that even at this age he was highly critical of their design.

Despite the serious misgivings of his family Whitworth quit his safe position in cotton in favour of the unfashionable craft of mechanic. He began his engineering career in Manchester, but his search for the best training inevitably took him to London to work for the firm of Maudslay, Sons and Field. His experience here under the eye of the famous Maudslay shaped the pattern of his life, and to some extent he can be said to have carried on where Maudslay left off. Whitworth worked for two other employers, the second of whom was Clement, whose contributions to the lathe and planing machine have been discussed in chapter 4.

In 1833 Whitworth returned to Manchester, renting a small section of a mill in order to set up business in his own right. He was the first person to concentrate on tool-making as a specific sphere of manufacture, and over his door was placed a sign for all to see, 'Joseph Whitworth, Toolmaker from London'. The choice of Manchester as the centre of his activities was a shrewd one. The northern city was near to sources of raw material, whilst the cotton mills and the new railways established in the area ensured that work would always be available. In time Whitworth was able to move from his rather cramped original premises, and by 1844 he owned a large mill in Chorlton Street where nearly two hundred men were employed.

Throughout his life in Manchester, Whitworth was the spearhead of a movement to refine and improve existing engineering methods. Until he made his influence felt, machine tools reflected the tradition of the eighteenth century clockmakers, where elegance of proportion was a major consideration. In some cases lathes were even styled with classical or baroque features like a piece of furniture. Whitworth changed all this, for to obtain the precision he required it was necessary to place the emphasis on sturdy machines such as his box-section lathe which could stand up to rigorous handling.

By 1849 he had taken out fifteen patents for improvements to machine tools, and although he lacked Maudslay's originality he invariably improved the design of whatever he manufactured. His modifications to machine tools included a shaping machine with quick-return motion, and in 1835 he designed his box-section lathe which was the first to possess both longitudinal and cross feeds driven from a single leadscrew.

Whitworth, through the production of these tools and also by advocating mechanisation of labour wherever possible, was helping to bring into practice what is today known as production engineering. He always tried to replace hand labour by machines. In his own factory in Manchester tools were placed together whenever their operations were related, and everything was carefully ordered even to the extent of having uniform paving stones to avoid unnecessary measuring whenever new machines were installed.

When Whitworth arrived in Manchester he found to his dismay that the standards of accuracy set by Maudslay had not spread throughout the industry. Craftsmen were still convinced that to work to the nearest sixteenth of an inch was quite acceptable. The campaign for accuracy rested squarely on the importance which Whitworth had been taught to attach to plane surfaces. In papers and in addresses to learned bodies he hammered away at this point, so that by 1856 the surface plates which were produced at his factory were at last coming into general use.

Two additional devices made by Whitworth considerably aided the dissemination of accurate standards. Both were finely controlled measuring machines depending on end measurement. The original model was accurate to a thousandth of an inch, and then in 1856 at the Institution of Mechanical Engineers, he proudly exhibited his machine which could detect differences of length as small as a millionth of an inch. It was the inventor's work on standardising the screwthread which was to make his name known throughout the world. As we have mentioned in chapter 4, Maudslay had been the pioneer in this work, but his efforts were not widely known. Furthermore, after Maudslay's death there was little possibility of gaining adherence to any common standard of thread in the free-and-easy world of mid-nineteenth century industry.

With considerable acumen, Whitworth decided to take a standard compiled by taking the average pitch and depth of thread made by leading British manufacturers. He then proceeded to sell the idea to others, and after publishing an influential paper on the subject in 1841 his efforts met with more and more success. In Whitworth's table each different diameter of screw had a definite pitch, and there was a constant angle of fifty-five degrees between the flanks of the thread. The inventor was also interested in regularising the gauges of wire and in the adoption of standard gauges throughout engineering.

This progress indicated that Britain was moving slowly but positively towards the acceptance of standard parts in machinery. Nevertheless, as Whitworth pointed out, on a number of important points the country was beginning to lag behind the United States. In 1856 he paid an official visit to the New World as a member of a commission which was to report on American industry. His findings were intended to be a warning to complacent British manufacturers. He emphasised that in the United States the best talent was eagerly conscripted into business, and that there was much greater readiness to accept machinery 'in every department of industry'. On his return from America, Whitworth was invited by the government to design a plant for the manufacture of rifles. This began his association with the armaments industry which eventually led to the foundation of the giant Armstrong-Whitworth gun factory. Whitworth himself designed a rifle which had no superior in range, accuracy, and in the degree of penetration achieved by its bullets. Characteristically, this fine weapon was rejected by the War Office, but within thirty years its revolutionary features were being copied by leading rifle makers throughout Europe.

Although nineteenth-century Britain lived by its manufactures, there was a notable reluctance to promote engineering and applied science as serious academic subjects. During the latter half of his life as a man of wealth and influence, Whitworth joined the battle to promote engineering education. He was largely responsible for establishing the first chair of engineering at Owen's College, Manchester, and his scheme of Whitworth scholarships gave young engineers of promise a chance of furthering their education.

In 1874 Whitworth turned part of his business into a joint stock company, making over a majority interest to his foremen and other employees. This was positive proof of the responsibility he felt for his workpeople, and it also demonstrated the important social benefits which he believed should derive from industry. The inventive spirit of the age as he saw it, should have the effect of 'emancipating men from the rude forms of labour, and making what were luxuries of one age to be looked upon in the next as the necessary conditions of existence'.

James Nasmyth, 1808-91

James Nasmyth was born into an interesting and influential Scottish family who lived in Edinburgh. His father, Alexander Nasmyth, was a well-known artist who included among his friends Robert Burns, Sir Walter Scott, and James Watt; he was also interested in engineering, and possessed a well-equipped workshop which had a treadle lathe. James, the tenth of eleven children, inherited his father's artistic talents and he had access to the workshop from childhood, quickly becoming skilful in the use of its equipment. On the lathe he manufactured spinning tops which he sold to his school friends, and he conducted a lively trade manufacturing small brass cannon and tinder boxes. He became a familiar figure in the engineering workshops of Edinburgh, where the friends he made were willing to teach him the secrets of their craft. At the age of seventeen he made a model of a beam engine which performed useful work in grinding colours for his father's paints. Nasmyth attended lectures at the university, not in engineering of course, for this subject was not taught at that time, and in the evening he studied at the School of Art. His main interests, however, were centred on his engineering work and he was soon building and selling model engines at £10 each. So keen was he that he set up a small brass foundry in his bedroom where he worked at night without disturbing the family. When he was nineteen he built a steam carriage at the works of a friend, which ran successfully for some months.

Through his contacts in the Scottish engineering firms he learned of the reputation of Maudslay, Sons and Field, and he was determined to go to London to obtain employment under Maudslay. In 1829 he journeyed to London, accompanied by his father, for an interview with the famous engineer. Maudslay had at that time decided against employing gentlemen apprentices, but he was so impressed by the models and drawing which Nasmyth had brought with him that he engaged him as his personal assistant. For two years Nasmyth worked alongside Maudslay in the master's private workshop where he absorbed the skill and enthusiasm of his tutor. He had the opportunity of meeting and listening to the discussions of the eminent scientists and engineers who visited Maudslay. During his time with the firm Nasmyth designed his miller for machining hexagonal faces of special collar nuts (see chapter 4), a flexible drive shaft, and an oil-heated stewpan designed to help his budget.

Shortly after Maudslay's death in 1831 Nasmyth left London to set up as a manufacturer of machine tools on his own account. He took with him a set of castings, which, through the kindness of Joshua Field, he had been allowed to make. From these castings he constructed the machines with which to start his factory, they included a lathe, a planer, and boring and drilling machines. This task was accomplished in Edinburgh where he was assisted by a skilled mechanic, Archie Terry, who was to remain with Nasmyth throughout his working life.

Premises were eventually found in Manchester where Nasmyth installed his machines and commenced business. Manchester proved to be a fortunate choice as the recent opening of the Liverpool and Manchester Railway had created a demand for machine tools. Nasmyth prospered to the extent that after five years he was obliged to seek larger premises, which he found at Patricroft, a few miles outside the city, and the new works were called the Bridgewater Foundry.

In 1836 he invented the shaping machine (see chapter 4), and installed in the foundry a mobile ladle operated by worm gearing. Nasmyth refused to take out a patent on this last device which was to eliminate a major source of foundry accidents.

Throughout his life Nasmyth kept a notebook in which he sketched details of ideas as they occurred to him. It is from this notebook that we have the original sketch of his most noteworthy invention—the steam hammer. The sketch originated when Nasmyth was asked for advice on the forging of the paddle shaft for a new steamship, *The Great Britain*. The shaft was too large to be forged by the conventional trip hammer. Nasmyth tells us that he conceived the invention within the space of half-an-hour, realising that the solution to the problem was to lift a heavy block vertically above the forging and then to release it at a determined point. He saw that the raising of the block could be accomplished by using steam to push a piston up a cylinder, and that the descent could be controlled by a lever which would release the steam through a slide valve.

The project of the paddle shaft was, however, dropped in favour of screw engines, and Nasmyth put his idea on one side. A few years later he visited the Creuzot works in France and was astonished to find a steam hammer identical to his design in use. On enquiring he was told that the design had been copied when a deputation from the French firm had visited Manchester during Nasmyth's absence. Fortunately, he was able to establish his rights to the invention and was granted a patent in 1842.

The steam hammer was adopted throughout the world as it greatly increased the size of the forgings that could be produced. It was soon improved by making the engine double-acting, steam being admitted above the piston to increase the power of the downward stroke.

With this invention Nasmyth's fortune was assured. He used the principle of the steam hammer in the steam pile-driver which he patented in 1843; this machine greatly reduced the time taken by the old manual methods. His flow of inventions continued, and included a keyway slotting machine, a hydraulic punch, a ball and socket joint, a grinding machine, and a ventilating fan for use in coal mines which replaced the primitive air pump of John Buddle (1807). Many years previously he had been impressed by the practice adopted at the Soho works where each machine was driven by its own vacuum engine. Now in his own establishment he applied the idea, using small individual steam engines to power the different machines. In doing this Nasmyth was anticipating the twentieth century use of motorised machine tools.

In 1853 he was appointed a member of the Small Arms Commission which recommended the adoption of the 'American System' for the production of the Enfield rifle.

Ill-health, brought about by overwork, forced Nasmyth to retire from business at the comparatively early age of forty-eight. For the remaining thirty-four years of his life he lived in retirement in Kent, where he was able to pursue his interest in astronomy. Into the short period of his working life Nasmyth had crowded some of the most influential discoveries of the nineteenth century.

Brown and Sharpe—Prototype of the American business corporation

National bias rather encourages us to ignore any contributions made by the inventors of other countries. Indeed, perhaps we venerate the inventor too much, and deprecate the contribution made by industrial firms to technical progress. Both these injustices are reason enough for choosing as the topic of this biography the American firm of Brown and Sharpe which operated during a dynamic period when the centre of gravity in industry was moving from Great Britain to the U.S.A.

The firm began in a small way in New England which had been an early centre of industrial activity. It originated in 1833 for the repair of clocks and mathematical instruments. In 1841, after a certain amount of quiet progress, the founder, David Brown, was succeeded by his more enterprising son, Joseph E. Brown. Then in 1853, with expansion in the offing, Lucian Sharpe was taken into partnership and the title J. R. Brown and Sharpe first appeared. The firm's initial reputation rested on the manufacture of reliable measuring devices at a moderate price. In 1850 Brown had designed a machine which automatically calibrated rulers with complete accuracy. Very soon afterwards he turned his attention to the manufacture of Vernier calipers.

At a competitive price such a tool was a great boon, and the calipers were described as 'the first practical tool which could be sold in any country at a price within the reach of the ordinary machinist'. Like Whitworth in Britain, it was natural that Brown and Sharpe should branch out from precision measurement into the movement for standardisation. This was the province of Lucian Sharpe who badgered fellow manufacturers, and finally in 1857 all the American companies, into accepting a standard American wire gauge. Having achieved this much, Sharpe drove home the idea of similar standards for sheet metal and other such products which would be made up into manufactured components.

In 1867 the company was approached by the superintendent of a local sheet brass company, S. R. Willmot, who had designed a micrometer. The micrometer had been invented in 1637, but its use was not yet widespread in the United States, and Brown and Sharpe were not enamoured with the idea of manufacturing Willmot's model. They were too enterprising to dismiss the idea altogether, and in August of the same year they saw a micrometer of a different type on display—the French Palmer micrometer. The result was a Brown and Sharpe model which combined the best features of both models. Palmer's method of clear lineal marking which was adopted is similar to that used today. The small compact shape of the Willmot type was retained, and there were certain additional features such as clamp for locking and an adjusting screw to compensate for wear.

The company's venture into machine tool manufacture began at the time of the American Civil War. There was intense pressure on armament factories, and consequently rich rewards lay in store for any designer who could improve gun-making machinery. Brown and Sharpe were initially engaged by the superintendent of the Providence Tool Company to aid him in retooling for the production of precision lock muskets. The commission was no ordinary one for the superintendent of Providence was F. W. Howe, the leading American expert in machine tools. He had worked previously with the famous company of Robbins and Lawrence at a time when they had brought out the turret lathe. He was also instrumental in the design of the Lincoln miller, and his association with Brown and Sharpe which lasted until 1873 was extremely beneficial to them.

At Providence an important stage in the manufacture of the muskets was performed by twist drills, and these had to be laboriously filed out by hand. In 1862, Howe suggested to J. R. Brown that a machine might be designed to do this work. The idea had an additional appeal, for a subsidiary sewing machine company of Brown and Sharpe also had need of drills. The outcome of the suggestion was the universal miller, which is described in detail on page 43 and which was undoubtedly an important development in the history of machine tools. The discovery made it possible to turn out quantities of twist drills capable of fast and precise work, which in itself was an important addition to engineering technique. The universal miller could also cope with a number of machining operations such as left-hand spirals and straight flutes in taper reamers. It was also quite simple to adapt it for straighforward milling work.

One of the subsidiaries of Brown and Sharpe maintained the company's original interest in clocks and watches, and it was this factory which focused J. R. Brown's attention on gear cutting, and eventually, in 1864, on the shape of the milling cutter. His formed milling cutter was once again a significant contribution to machine tool development.

The more exacting demands for accurate components which could be finished after hardening had created a great interest in precision grinding. Brown and Sharpe first began making grinding machines to hasten the manufacture of sewing machine parts for which they were under contract. In 1867 their first attempt consisted of a powered grinding wheel mounted on the carriage of a modified lathe. The success of the universal miller, however, suggested that something along the same lines might be possible with grinding. The original idea of a universal grinder probably originated in 1868, but the project remained in abeyance and was revived again in 1873 when a vitreous emery grinding wheel came onto the market. The machine finally appeared in 1876, shortly after J. R. Brown's death in July of that year. It was a radical improvement on previous models. In particular, the work traversed past the grinding wheel, thus following the mode of operation of the milling machine rather than the lathe. A protegé of J. R. Brown, H. M. Leland, who became a famous president of the Cadillac Motor Corporation, considered the universal grinder to be Brown's most important achievement. 'In developing and designing this machine', he wrote. 'he stepped out into entirely new ground, and developed a machine which has enabled us to harden our work first, and then grind it with the utmost accuracy.'

Thomas Newcomen, 1663-1729

Thomas Newcomen was born at Dartmouth in Devon where, after serving an apprenticeship, he started his own business as an ironmonger and blacksmith. In the neighbouring county of Cornwall there were a large number of tin mines, and mining, particularly for coal, was becoming increasingly important throughout Britain. Although we know relatively few details about Newcomen's life, we believe that it was through his connections as an ironmonger that he became aware of the problems involved in draining the deep Cornish tin mines. It appears likely that Newcomen had started experimenting with the use of steam to solve these problems by 1698.

Unfortunately for Newcomen a similar idea had occurred to Captain Thomas Savery, a military engineer who was to become a Fellow of the fashionable Royal Society. Savery was in a better position to exploit his idea and, although his steam pump (see chapter 5) proved incapable of solving the mine drainage problem, his influence at Court helped to obtain for him a patent. This patent covered any method of 'raising water by the impellent force of fire' and was to last for the ridiculously long term of thirty-five years. Its existence hampered Newcomen throughout his life, and he was forced to reach a financial agreement with Savery before he could market his engine.

There is little doubt that Newcomen, and his assistant, John Calley, experienced many setbacks before the engine worked to their satisfaction. The biggest problem was that of condensing the steam inside the cylinder to create a vacuum, and no doubt it took many years of patient experiments before the eventual method of water injection was perfected. It was not until 1712 that the first Newcomen engine was installed, not in Cornwall as might have been expected, but near Dudley Castle in the Midlands. This engine made twelve strokes a minute, lifting ten gallons of water through 150 feet at each stroke.

In a contemporary print of the Dudley Castle engine Captain Savery and Newcomen are described as the joint inventors, evidence of the confusion brought about by the patent difficulty. Even after Savery's death in 1715 control did not pass to Newcomen, for the patent rights were acquired by a syndicate which was to draw substantial royalties on the engine until the expiry of the patent in 1733. Neither did Newcomen receive the acclaim due to him, for it was difficult for the eminent scientists of the time to allow that a country ironmonger could conceive such a fundamental invention. Thus we find Desaguliers, a distinguished contemporary scientist, in his first written account of the engine suggesting that Newcomen had relied heavily on the work of Papin and Savery, and that anything further he had achieved could only have been by lucky chance.

Undoubtedly the engine had many faults, not least the excessive coal consumption which inhibited its use in the Cornish tin mines. The engine was costly to install, figures upwards of £1500 being quoted, and the use of brass for the cylinder casting not only added to the cost but limited the diameter of cylinder that could be produced.

After the expiry of the patent in 1733 the popularity of the engine increased, and with the availability of larger cast iron cylinders from Coalbrookdale it proved invaluable to the owners of coal mines who were driving deeper shafts as the demand for coal increased. By 1769 ninety-nine Newcomen engines were counted in the North of England alone, and they had been installed as far afield as Russia, Austria, and Hungary.

L. T. C. Rolt summarises Newcomen's contribution to engineering with the following comment: 'Crude, inefficient and slow though his engine was, he was the first man to harness power in a cylinder with practical success. In this sense Newcomen was the great innovator, and all who came after him, even James Watt, were no more than great improvers.'

James Watt, 1736-1819

Although the Newcomen engine came as a boon to mine-owners in the early stages of the Industrial Revolution, by the second half of the eighteenth century the limitations of the Newcomen engine were frustrating to industrialists who had urgent need for a source of power more reliable than the water-wheel.

The man who achieved the necessary fundamental improvements to the Newcomen engine was James Watt, born at Greenock, near Glasgow, in 1736. Watt came from a respectable family; his grandfather had been a teacher; and his father, a councillor and magistrate, was a trained craftsman who ran his own business, specialising in nautical instruments.

James Watt was a sickly child, and suffered from migraine throughout his life; it was decided that he should train as an instrument maker rather than enter his father's declining business. He spent a year in London learning the craft, and on returning to Glasgow obtained permission to set up as instrument maker to the University with a shop in the precincts. Watt easily made friends with a number of the academic staff, and sometime in 1764 a Dr Anderson brought to him for repair a model of a Newcomen engine belonging to the University. Watt repaired the engine but had difficulty in keeping it running, and he sought advice from Dr Black, the distinguished chemist, on the nature of the scientific problems involved. Eventually Watt concluded that the heat losses in the cylinder walls, magnified by the 'scale effect' of the model, were the cause of the trouble. He turned the problem over in his mind until, on a Sunday afternoon in May 1765, the idea of a separate condenser (see chapter 5) occurred to him while strolling across Glasgow Green. Very soon he had constructed a model which proved that this theory was sound.

Watt had a shrewd idea of the value of his invention but he needed to build a full-size engine to test it, and that required financial backing. Once again Dr Black proved useful by obtaining an industrial sponsor for Watt in the person of Dr Roebuck of the Carron Iron Foundry, a man with large industrial and mining connections in Scotland and Birmingham. Watt struggled to perfect his engine while working as a surveyor to support his family, but he made little progress in the next few years. He obtained his historic patent for 'A New Method of Lessening the Consumption of Steam and Fuel in Fire Engines' in 1769.

Ironically, it was Roebuck's bankruptcy which was the turning point in Watt's career, for the former's share in the patent was bought by Matthew Boulton of Birmingham. Boulton had a flair for business, a keen advertising sense, and wide contacts. Not least, he was the owner of the Soho factory in Birmingham, where he had available skilled labour, good machinery, and excellent facilities for making the engines.

Watt moved south in 1775 and in the same year gained, by parliamentary petition, an extension of his patent for an additional twenty-five years. This gave the firm of Boulton and Watt a monopoly on the new engine until 1800, and they proved most adept at defending their patent in a succession of legal actions. Watt wished to spend time experimenting with the engine which he had brought down from Scotland, but this was hardly consistent with Boulton's purpose. He lost no time in hustling Watt into designing two massive engines, the larger of which had a cylinder diameter of 1.2 m.

The success of Watt's engine depended upon the cylinder being bored accurately, and fortunately John Wilkinson's cannon-boring machine was available. A description of this early machine tool can be found in chapter 4.

Fig. 93 A double acting Watt engine in use until 1840

The casting and boring of the cylinders and condensers were undertaken by Wilkinson, and were generally transported directly to the clients.

Times were propitious for an improved steam engine and it was soon reported that 'London, Manchester, and Birmingham are steam mill mad'. The first engine was sold to John Wilkinson to work the blast at Broseley, and soon a steady demand came from mine-owners, ironmasters, canal-owners, and the like, who found that the new engines were superior to the Newcomen type. However Watt and Boulton drove a hard bargain, as indeed Newcomen had before. They insisted on a form of contract by which they received a third share of all the fuel saved. On such terms some bargaining might have been expected, but Watt's attitude was very much 'take it or leave it', and after visiting a company at Shadwell he wrote: 'I told them we did not solicit their orders, and would wait patiently until they were convinced.' It certainly says a good deal for the new engine that so many businessmen were prepared to accept it on these terms.

After a hectic intial period during which he spent some time in Cornwall erecting engines, Watt was largely free from worries on the business side which were attended to by his partner. This allowed him to follow up his real interest which was in technical development and design. He remarried, and moved with his new wife, Anne Macgregor, to a larger house in Harper's Hill in Birmingham, where he had his own drawing office at home. His inventions by no means ceased with the condenser, and we must briefly consider the major improvements made by him to the steam engine. Some were made before he settled at Harper's Hill, but most were incorporated in the model of the steam engine built in 1788.

1. The open-ended water-sealed cylinder of the Newcomen engine was discarded and the cylinder totally enclosed in a steam jacket to keep it hot.
2. A stuffing box was provided to prevent the escape of steam through the piston rod aperture.
3. Steam at slightly above atmospheric pressure produced the power stroke.
4. Condensation of the steam to produce the vacuum took place in the separate condenser which was evacuated by a pump operated from the beam.

Boulton was anxious to produce the engine as a source of rotary power for use in mills and factories, and to this end Watt incorporated the following improvements:

5. The engine was made double-acting, that is, the vacuum was created on alternate sides of the piston so that each stroke became a power stroke.
6. This entailed a rigid connection between the piston rod and the beam arch, and to keep the end of the rod moving in a straight line Watt designed his famous linkage.
7. Rotary motion was produced through a sun and planet gear.
8. To cater for variable loading Watt adapted the centrifugal governor to control the steam supply automatically.
9. He invented the indicator to measure cylinder pressures, and he devised a unit of work—the horse power—in terms of which he expressed the output of his engines.

For all its advantages, the Watt engine worked on a steam pressure which was little above that of the atmosphere, and the only hope of substantially increasing the output lay in the development of the compact high-pressure engine. Watt firmly believed the use of high-pressure steam to be dangerous and impracticable, and we see him in his later years resisting testily progressive engineers like Hornblower and Trevithick, the pioneers of high-pressure steam.

The inventor's life, though successful, was dogged by ill-health. He suffered a crippling number of bereavements amongst his family, even for the eighteenth century. Life made him dour and cautious, but his mental agility won him wide respect. He was elected a Fellow of the Royal Society, and he was an active member of the Birmingham Lunar Society to which some of the most brilliant figures of his age belonged.

Richard Trevithick, 1771-1833

Trevithick epitomises the popular and romantic idea of an inventor. He possessed an original and ingenious mechanical talent and an almost boundless fund of vitality, but, unfortunately, he lacked the business sense to see his ideas through to a successful conclusion.

Born in 1771 near Carn Brea in Cornwall, Trevithick grew up during the years in which the steam engine was really coming into its own in the tin mines of the county. At the age of nineteen he was being consulted by mine-owners on steam engines, and even at this age he seemed to have established himself as a 'character'. At Crane's mine, for example, on one occasion he is reported to have thrown a smith's sledgehammer over the engine house roof.

Trevithick's early experience had familiarised him with the technique of compounding the Watt steam engine, and this was to give him a lifelong enthusiasm for the potentialities of high-pressure steam. As Watt had done before him, he took scientific advice from a friend, and was assured that it was perfectly feasible to drive the piston using high-pressure steam without a vacuum.

Trevithick, working on this theory, built his first compact engine which he patented in 1802. One of its most important features was the boiler, which had to be capable of withstanding steam pressure of at least fifty pounds per square inch (3.5 kg/cm^2). He experimented with a number of shapes, but at last arrived at the cylindrical boiler, which he was to develop later into a significant discovery in itself.

The boiler had an internal U-shaped flue, and the cylinder was sunk into the boiler itself in order to keep it hot. There was also a splendidly simple and direct method of distributing steam to the cylinder and of connecting the piston rod to the crankshaft.

An engine was now available which was powerful and compact, far more versatile than the cumbersome monster of Boulton and Watt. Unknown to Trevithick, similar work on high-pressure steam was being undertaken by Oliver Evans in America, but it was Trevithick alone who had the breadth of vision to exploit high-pressure steam for the purposes of transportation.

In 1801 Trevithick built a steam carriage to travel on the roads. With perseverance, the initial running faults might have been remedied, but after two characteristic accidents, one of which resulted in the machine tearing down the iron railings around the house of a rather peppery London gentleman, the inventor soon lost interest. Instead he took up a new venture, for which he is today most remembered—the steam locomotive designed to run upon a railway. He built a locomotive which made the famous run between Merthyr and Abercynon near Cardiff in 1804. Although this journey has been generally accepted as the first to be made by a steam locomotive, there is now some reason to believe that an earlier model was built by Trevithick to run at Coalbrookdale as early as 1802. The firm of Boulton and Watt were, not unnaturally, disturbed by the turn of events, and since they steadfastly refused to adopt a progressive outlook on high-pressure steam, they felt obliged to belittle Trevithick's work in any way possible. They fastened in particular on an explosion of a Trevithick boiler which was caused by a labourer who had fastened down the safety valve whilst he was having his break. Trevithick's answer to such hazards was the safety plug—a lead rivet which was so placed that it melted when the water in the boiler became too low, so allowing the steam to escape without doing damage.

Inventive genius was no guarantee of success in Trevithick's case, and in 1811 his financial difficulties forced him to sell out his patent rights completely and declare himself bankrupt. He returned to Cornwall, the source of his first inspiration, and became engaged in a number of schemes, all involving the use of high-pressure steam. Probably the most important of Trevithick's work in these years was the development of the Cornish engine—a beam engine similar to that of Watt, but using high-pressure steam.

In 1812 Trevithick built a boiler of an entirely new type. This was the so-called Cornish boiler, destined to be a most efficient method of raising steam.

The fire was placed in a cylindrical tube which was enclosed by the boiler itself so that the hot flue gases could provide the source of heat.

In 1815 Trevithick built an incredible contraption called the whirling machine, which it was commonly believed had been constructed to throw cannon balls at the forces of Napoleon. The steam supplied from the boiler at 100 lb per inch square (7 kg/cm^2), rushed through a hollow axle and was expelled from two nozzles placed at the extremities of the whirling arms. Since the nozzles faced in opposite directions, it can be seen that the pressure of the steam would have caused the arms to rotate at high speed. Trevithick could not achieve anything approaching the necessary velocity of steam, and he dropped the project with his usual impatience.

As if his life had not been interesting enough, Trevithick spent eleven vital years from 1816 to 1827 pursuing his fortunes in the turbulent states of South America. He went out in the hope of applying his high-pressure engines in the high altitude of the silver mines of the Andes. He began hopefully, but his prospects were ruined by the outbreak of the South American Wars of Independence, and his life became a series of amazing adventures. He came home on borrowed money, and returned to the West Country to find that in the years he had been away others had made use of his ideas. Under less imaginative but more careful guidance than Trevithick's, his ideas prospered.

In many ways Trevithick suffered because his theories were in advance of the materials and skills of his time, but had he not gone off to South America, there is little doubt that greater progress would have been made. With all his limitations, Trevithick was an engineer of genius, and his stationary and locomotive engines were at the very root of nineteenth-century prosperity and expansion.

The Hon. Charles Algernon Parsons, 1854-1931

The steam turbine had existed as a kind of engineering plaything since the time of Hero, the Greek, but it was not until the latter part of the nineteenth century that it became a practical proposition. The man mainly responsible for this achievement was Charles Algernon Parsons, one of the rare examples of the nineteenth-century engineer born into the upper class. As the youngest son of the Earl of Rosse he enjoyed a number of advantages in his early life, including independent means and a favoured apprenticeship. Parsons also had the opportunity of reading for a degree in mathematics at Cambridge, while his father, who had been at one time President of the Royal Society, helped to stimulate his taste for science. Such a favoured background was important because the development of the steam turbine was not a matter which could have been solved by practical talents alone.

Parsons began work on the steam turbine when he was still supposed to be learning the practical side of engineering at Armstrong's works in Newcastle upon Tyne. He was influenced by the example of the water turbine in his attempt to produce a machine which would turn an electric generator, also of his own design. The steam turbine patented by Parsons on 23 April 1884 is described in detail in chapter 5, and it undoubtedly marked a turning point in the evolution of steam power and in the history of electrical engineering. It will be remembered that the steam moved axially through a set of turbine blades, causing the rotor shaft to move at high speed within its cylindrical casing. The blades were so arranged that the fall in steam pressure occurred progressively along the whole length of the turbine. Subsequently Parsons experimented widely on the shape of the turbine blades themselves so that he could obtain the most efficient performance.

The original turbine achieved a speed of 18,000 r.p.m., and a blade velocity of 250 75 m per second, which was capable of turning a dynamo producing 75 amps of direct current at 100 volts. Parsons also anticipated a modern system of lubrication by adding a direct-drive screw pump which circulated oil to the turbine's main bearings. He had his difficulties, and for a while he lost his patent rights, which remained under the control of his former employers. This meant that he had to continue his experiments on the radial or outward-flow turbine which was not so efficient as the axial-flow type. Fortunately, in 1893 he was able to buy back his full patent rights for the modest outlay of £1500, and by that year he was also securely established in his own firm and in a position to exploit his discovery to the full.

To lessen steam consumption Parsons added a condenser to the turbine, which also had the effect of reducing engine vibration. The result was a power unit which was eminently successful in its prime function of driving electric generators. A tandem cylinder turbine like the one Parsons built for the German town of Elberfeld, for example, provided the cheaper electricity which helped to change the whole pattern of life at the beginning of this century.

Parsons saw that the steam turbine could be easily applied to the propulsion of ships. This was at a time when naval power was regarded as a matter of great concern, and with a beautifully calculated sense of daring, Parsons decided to demonstrate his vessel *Turbinia* at Queen Victoria's Jubilee Naval Review in 1897.

As the great battle fleet was drawn up in formal array, the 44-tonne vessel, powered by a 1500 kw turbine, burst through. The 'captains and kings' were startled by her manoeuvrability and her amazing turn of speed. At a maximum speed of $34\frac{1}{2}$ knots *Turbinia* was able to outpace the fastest destroyers which at that time could only make 27 knots.

Parsons had made his point, and his invention was quickly followed up in naval circles. It was found, however, that the turbine speed was too fast for the screws, and in 1910 Parsons designed a speed-reducing gear which allowed the turbine to run at its full speed and efficiency.

Parsons lived to become a 'Grand Old Man' of British engineering, and when he died in 1931 it was recognised that his famous turbine patent no. 6735 stood, with Watt's condenser, as one of the really great engineering discoveries.

George Stephenson, 1781-1848

The lives of George Stephenson and his son Robert belong more especially to the history of transport. Yet it is appropriate that both should find some place in a study of mechanical engineering. The father was the first president of the

Institution of Mechanical Engineers, and both father and son were passionately concerned with the steam engine in its most relevant application to their own times—locomotion.

George Stephenson's birth was obscure enough; he was the son of a steam engine fireman and was born in the Northumberland mining village of Killingworth. Yet in their own way the circumstances of his upbringing were propitious, especially the fact that a clumsy wooden railroad ran past his house. This was a means of transport quietly gaining favour over the canals which were expensive to construct. Fortunately again for Stephenson, the local mine-owners used steam power quite extensively, and during his early years he was able to get an excellent working knowledge of the steam engine in his position of fireman, and later, at the age of seventeen, as an engineman.

Up to this point in his career he was illiterate, but at the cost of fourpence a week he taught himself to read, write, and do basic arithmetic. Stephenson was undoubtedly hampered all his life by his lack of knowledge of engineering theory. His mark was to be made by dint of a native mechanical genius, and in 1811 he got his first real opportunity. A Smeaton engine which had been installed in the Killingworth colliery broke down, and where others had failed he diagnosed the trouble and quickly set it to work. The powerful Grand Allies group of pit owners were impressed, and Stephenson was offered the position of chief mechanic, with freedom to undertake work for other local mine-owners.

Once established, Stephenson's ability and business shrewdness took him far. He became extremly interested in the enterprising example of William Hedley of Wylam on the north bank of the Tyne, who had introduced Trevithick's idea of a steam locomotive to north-eastern England. The lack of horse fodder had reached serious proportions because of the war against Napoleon, and this enabled Stephenson to win the support of the Grand Allies Co. to his scheme for a railway. George Stephenson may not deserve the title 'Father of the Railways'—this undoubtedly belongs to Trevithick—but the forceful northerner however produced in the *Blücher,* his first locomotive of 1814, a highly serviceable engine which could be used as the basis of future development. It incorporated the following features:

1. The drive from the pistons was transferred to the connecting rods by guided crossheads.
2. The exhaust from the cylinders was turned into the chimney through an upturned pipe.
3. Transmission was effected by coupling the connecting rods directly to crank pins on the wheels.
4. A chain drive connected the leading wheels with the rear axle.
5. A form of steam-actuated suspension enabled the locomotive to ride more easily.

In 1815 Stephenson won considerable prestige by inventing a miner's safety lamp, the 'Geordie lamp' as it became known. Unfortunately, Sir Humphry Davy, an eminent man of science who had also devised a safety lamp, refused to consider that a semi-illiterate mechanic could achieve such a feat. In a barrage of rather unsavoury publicity he did everything he could to discredit Stephenson's achievement.

In 1821 Stephenson launched himself into the world of civil engineering by undertaking to survey the first intertown railway in the world, the Stockton and Darlington. By 1823 he had become an important figure in the northern business world, and in association with a number of Quaker partners he set up a company for making locomotives. According to the partnership arrangements, the works at Forth Street in Newcastle, soon to be world famous, were to be managed by the young Robert Stephenson. From here, as we shall see in the biography of Robert Stephenson, were to emerge the *Rocket* and many other famous locomotives, for the Stephensons were now in a position to influence locomotive design for a long time to come.

George Stephenson was responsible for the construction of the Stockton and Darlington railway, and provided the rolling stock, but he now tended to concentrate on the surveying and construction of railways, leaving the building of locomotives to his son and others. In 1830, after the most resolute and bigoted opposition from landowners, canal owners, and other forms of vested interest, he completed the Liverpool to Manchester line. The engineering problems were tremendous. There was a great tunnel of 2 km to cut at the Liverpool end, but worst of all was the problem of crossing the Chat Moss bog at the Manchester end, a morass which was believed by many people to be bottomless. Stephenson overcame the problem, partly by draining the bog, and partly by building up an embankment on rafts of brushwood and heather.

Towards the end of his life George Stephenson's reputation amongst the railway engineers, so the tradition goes, was so considerable that, finding him excluded by his background from the Institution of Civil Engineers, they founded the organisation later to become the Institution of Mechanical Engineers, and invited him to be their first president. He died at his Chesterfield home, Tapton House, in August 1848. Stephenson was not an original genius of the measure of Trevithick, but his tremendous drive and innate faith in the potentiality of railways made him an outstanding figure of the nineteenth century.

Robert Stephenson, 1803-59

The early years of Robert Stephenson were completely dominated by a strong paternal influence consequent upon the death of his mother when he was one. The elder Stephenson held a lifelong contempt for theoretical engineers, but he made sure that his own son had a sound engineering training. Robert was sent to a good academy in Newcastle, and his schooling was supplemented by a course of technical reading. He would bring back to the family cottage at Killingworth books from Newcastle library and read them out to his father who was determined to profit from his investment in Robert's education. The latter part of his training was not unlike a modern engineering sandwich course; he served his apprenticeship under a colleague of his father, Nick Wood, and then under George Stephenson himself. Finally, to give him polish, he was sent for a term to Edinburgh University.

George Stephenson was certainly a rather brusque and tough parent, and he tended to exploit his son's theoretical knowledge to his own advantage, but at the same time he transmitted to his son his own passion for engineering. Moreover, as George Stephenson progressed, he took his son up with him, and when negotiating the foundation of a locomotive works at Forth Street, Newcastle, he was careful to have Robert appointed as a managing director at the age of twenty.

For three years, from 1824 to 1827, Robert cut himself adrift and went to South America in an obvious attempt to assert his independence. Like Trevithick before him, he went out to utilise high-pressure steam engines in the high altitude mines of Columbia. The experience was disheartening, but he returned more self-possessed and ready to cooperate on more equal terms with his father.

Back at the Forth Street Works in Newcastle, Robert settled down to a programme of locomotive design. He broke away from his father's conceptions, which had tended to look like a mobile adaptation of the heavy and cumbersome fixed steam engine. Robert used the experience of the steam carriage builders, Trevithick and Gurney. The fruits of his labour were a revolutionary series of locomotives beginning with the *Lancashire Witch* of 1828, and continuing with the *Rocket* of 1829, the *Northumbrian* (1830), and the *Planet* (1830). In October 1829, the directors of the Liverpool and Manchester Railway met to decide on the best type of locomotive to run on their line. George Stephenson was, of course, employed by the company, but this did not mean that the decision was cut and dried in favour of the Stephensons. A series of exhaustive trials was held at Rainhill, and before an audience of ten thousand every train was required to haul its load up and down the one-and-a-half mile course some twenty times. The *Rocket*, which was able to show her superiority over all the other entrants, had a number of new features:

Fig. 94 Stephenson's *Rocket*

1. The cylinders were inclined at an angle of forty-five degrees, and provided a direct drive to the big new driving wheels in the front.
2. A revolutionary boiler contained a number of small tubes through which the heat from the fire passed. This meant an increased heating surface, and consequently greater power.
3. A rotary plug valve in the main steam pipe controlled the admission of steam to the cylinders so that the expansive power of the steam was used to great effect and with greater economy of fuel.

The *Planet* built by Stephenson in 1830 has been described as 'the true prototype of the steam locomotive of modern times'. For the first time the cylinders were mounted beneath the engine, and again for the first time, attention was paid to safety. There were bearings for the cranked axle, and a stronger frame all round.

By the age of twenty-seven Robert Stephenson's reputation was such that he was able to move out into the more rewarding field of railway survey, bothering less with his locomotive construction. In 1830, the same year in which he built the *Planet*, he began the survey of the 180 km railway between London and Birmingham. It proved a most exacting project of civil engineering, involving cuttings through soft rock at Tring, and tunnels at Kilsby and Primrose Hill in London.

Stephenson's fame meant that his services as an engineering consultant were in demand all over the world. One of his great achievements was the Britannia bridge which, like many of his other bridges, was constructed from huge iron tubes. These weighed up to 1500 tonnes apiece; they were floated into position and then raised by hydraulic presses.

Like his friend, I. K. Brunel, Stephenson's hectic ubiquitous life ultimately injured his health. His constitution had never been robust, and the sheer volume of hard work precipitated his death in 1859 at the age of fifty-six. He was not only a great engineer, but thoroughly professional in a way in which his father with his humble background could not be. George Stephenson, for example, took technical innovations like Brunel's broad gauge, as a personal insult, but Robert was able to combine a keen rivalry between himself and Brunel with a warm friendship. Both Stephensons established their fame upon the steam engine. Robert understood the challenge and potential of this new force as much as his father: 'I will fight for them to the last', he once wrote of his locomotives, 'they are worthy of conflict'.

Isambard Kingdom Brunel, 1806-59

I. K. Brunel occupies a very special place amongst the engineers of the nineteenth century. He belonged to the heroic generation who practised before engineering became excessively specialised. He was born in 1806, the son of Marc Isambard Brunel, a French emigré from the Revolution. Brunel senior instituted, with Henry Maudslay, a method of making ships' pulleys which has been acclaimed the earliest example of mass production. His father's success in Britain assured I. K. Brunel of a good start, and he was well educated, spending some time in the Lycée Henri Quatre in Paris, a school famous for its mathematics. After this came a sound practical training with a famous French clock and instrument maker, followed by free access to the works of his father's former associate, Maudslay. It was there, as Brunel later wrote, 'I probably acquired all my early knowledge of mechanics'.

His first taste of responsibility came early, for in 1828 Brunel was obliged, because of his father's illness, to take charge of a large-scale project for a tunnel under the Thames. It proved a perilous assignment, and Brunel was nearly killed when the workings became flooded in a torrent of river water. But responsibility only whetted his appetite, and in the following year at the age of twenty-three he submitted a revolutionary design for a suspension bridge over the Avon Gorge at Bristol. The opposition to this rather callow young engineer was formidable, and included Telford himself but a suspension bridge, which was a new concept in Britain, obviously appealed to the cost-conscious promoters. After a series of rather underhand machinations which often accomplished such an appointment, Brunel's design was finally accepted, though owing to financial difficulties the bridge was not completed until 1864, five years after Brunel's death.

Brunel was fortunate to be associated with Bristol at this time, for the merchants of the port were concerned over the rivalry from Liverpool, and to invigorate trade they planned a railway link with London. It says a great deal that, at twenty-seven, Brunel's reputation was such that he was given the appointment of surveyor and engineer to this proposed Great Western

Railway. To railway enthusiasts in Britain the Great Western Railway, which in 1947 became the Western Region of British Railways, has always occupied a place of special affection. This in no small way is due to the efforts of Brunel because from the very outset he determined that everything should be on 'the grand scale'.

In course of time the plans for the line were extended to cover a link with the West Country and South Wales, but before any work could be started it was necessary to survey the land in the teeth of a strong vested interest from local landowners. Once the survey was completed the struggle moved to Parliament where the coaching and canal interests combined with the landowners in trying to defeat the Bill sanctioning the scheme. Brunel was cross-questioned for eleven days by a barrister, but, unlike George Stephenson on a previous occasion, he did not get befuddled and lose his temper. At last the Bill passed through Parliament and Brunel could devote himself to this great engineering challenge, which was to take him five-and-a-half years to complete.

He began in a typically forthright manner by challenging one of the sacrosanct laws of railway practice, the standard gauge of 4 feet $8\frac{1}{2}$ inches (1·43 m). The use of this had been arbitrarily set down by George Stephenson, who had found it used on the old colliery tramways. Brunel dismissed it out of hand and instituted the famous broad gauge of 7 feet (2.13 m), which assured fast speeds of up to sixty miles (96 km) an hour in safety. Not for the last time he was opposed for an allegedly extravagant idea, and with his future endangered by manoeuvrings both inside and outside the company, he backed down and abandoned the broad gauge against his better judgement. An Act of 1846 prohibited further building of broad gauge railways although the last of the track did not disappear until 1892.

Brunel was the fount of many other highly ingenious ideas. He rejected the existing solid iron rail as both too heavy and too costly, and replaced it with the bridge rail. It was hollow with an arched cross section which made for greater lightness, but not at the expense of strength.

When the G.W.R. link between Paddington and Bristol was finally opened in 1841, the project was estimated to have cost £6,500,000. There had been among other things great difficulties with the terrain, and between Bath and Bristol alone, it was necessary to build seven tunnels, four bridges, and a viaduct. Brunel's greatest headache was the Box tunnel, the longest tunnel bored until that time. It was nearly 3.5 km long, on a slight gradient, and its cost must be measured, not only in time and money, but in the lives of nearly a hundred navvies out of a working force of four thousand.

Apart from his famous bridges over the Tamar and over the Wye at Chepstow, Brunel's other memorable contribution to the railways was a noble failure. It was the famous atmospheric railway which he tried unsuccessfully between Exeter and Newton Abbot. In theory it was brilliant. A continuous pipe, fifteen inches in diameter, was laid between the rails, and the carriages were attached to it by means of a suspended piston. No locomotive was necessary for the carriages were propelled by suction when the air was evacuated from the pipe by means of stationary pumping engines placed at stages along the line. Unfortunately, rats and sea air attacked the leather flap which made the connection between pipe and piston, and the whole experiment had to be written off as a ghastly failure which had cost half a million pounds.

It was through his connection with the G.W.R. that Brunel gained financial support for his three great transatlantic ships, the *Great Western*, the *Great Britain*, and the *Great Eastern*, each successively more ambitious in tonnage and power.

One great undertaking upon another finally broke the engineer's health, and he died in 1859 leaving behind him a number of incomplete schemes and ideas which were never taken up because they were too much in advance of the times. One can certainly sympathise with the men who refused to risk capital in his imaginative projects, but at a turning point in engineering he brought an exhilaratingly fresh approach. One of his subordinates once spoke of Brunel's 'extreme and unprecedented insistence on excellence of work'. In this sense he represents a link between the new world of engineering and the high standards of men like Maudslay who had been one of his tutors.

Henry Ford, 1863-1947

The life of Henry Ford, founder of the giant motor corporation, forms a link between the nineteenth century and the twentieth. He transformed the motor car, a nineteenth-century invention for the wealthy, into something which could be made for the ordinary man. Furthermore, in order to do this, he organised the manufacture of cars so effectively that his production methods became a model for modern industry in general.

Ford was born on a farm near Detroit in the United States in 1863, and he had begun to tinker with farm machinery by the early age of ten. As a young man Ford drifted from being a watch repairer to working on gas and steam engines. Both were rather mundane jobs, but they gave him sound mechanical experience for the future. He built his first car in 1896, a fairly conventional model typical of the many being constructed in backyard garages throughout America and Europe. Like other car makers then and since, he realised that he could do himself a lot of good by racing, and in 1901 he won an important race in Michigan, gaining recognition for himself and his newly-formed company.

But Ford's vision was wider than those of other small car manufacturers; he wanted to bring out a reliable car which, as he put it, 'could replace the family horse'. He also had a genius for using what other people had to offer. In 1903, for example, he badly needed money, and was able to obtain it from his coal dealer, Alexander Malcolmson, who became his business manager.

The first of the famous line of Fords was produced in 1903, and was called the Model A. Ford adopted a practice already established by Ransome Olds, that of using standard parts, so that all his transmission units, for example, could be obtained from one company, and all his wheels from another. He also made it easy for Ford owners to have their cars serviced by setting up a nationwide network of trained mechanics who could undertake repairs and were ready to listen to complaints. At this point many men would have leaned back and enjoyed the fruits of success, but Ford was by no means ordinary.

By 1907 he had produced Models A, C, N, and R. He now ventured everything on his most ambitious gamble of all, Model T.

For the first time a car was completely enclosed so that it could be used both in summer and winter. Ford also added important accessories such as windscreen wipers without any extra charge. The transmission of the car was partly automatic, and both the ignition system and the carburettor were simple but effective. To start the car it was necessary to pull a small wire sticking out in front, and this controlled the mixture of air and gas which entered the carburettor. With the other hand the driver would need to turn the hand crank operating the flywheel magneto. The body itself rode high above the ground because of the primitive roads, and for added strength it was constructed of a new steel containing vanadium. The Ford T was a phenomenal success, and the final output for the model achieved the figure of fifteen million.

A clever advertising campaign to which Ford paid great attention resulted in a record demand right from the outset. It was quite impossible to meet this with the existing factory set-up, and so Ford, in cooperation with his assistant, Walter Flanders, inaugurated an entirely new system which has become known as mass production.

One afternoon in August 1912 the first experimental assembly went into operation. A bare automobile frame was pulled slowly forward by a rope revolving round a drum. As the chassis moved down the line, parts were added by six assemblers, and in the final stage the engine was lowered into place by chains. Even such a rudimentary production line cut time by more than half, and the operation was quickly streamlined. Very soon the work of the Ford plant became geared to supplying the main assembly line which had smaller lines moving in to it. Raw materials were delivered according to a strict schedule, and parts for the car were designed with one prominent idea in mind—speed and economy of production.

Henry Ford is well remembered for the remark he once addressed to his customers. He told them that they could buy a 'car of any colour so long as it was black'. There was a good deal of shrewdness behind this apparently chance comment, for black paint gave the Ford car a fine surface finish.

Ford did more than any other manufacturer to bring the motor car within the budget of the ordinary man, and his techniques of mass production were adopted in many other firms where efficiency was held to be important.

Additional Reading

Engineering Heritage, 2 vols, Institution of Mechanical Engineers 1963, 1965.

L. T. C. ROLT, *Brunel*, Longman 1960.

——, *George and Robert Stephenson*, Longman 1960.

——, *Great Engineers*, Bell 1962.

——, *Victorian Engineering*, Allen Lane 1970.

D. S. LANDES, *The Unbound Prometheus*, Cambridge University Press 1969.

S. D. CHAPMAN, J. D. CHAMBERS and T. R. SHARPE, *The Beginnings of Industrial Britain,* University Tutorial Press 1970.

Examination Questions

1. *Either*
   (a) Describe the furnace and process used for smelting cast iron, mentioning the raw materials and their function.
   Or
   (b) Describe the part played by the Siemens brothers in the development of the steel industry. What is the name of their process?
2. The following are all names associated with the development of the screw-cutting centre lathe: Roberts, Fox, Ramsden.
   Give details of the contributions made by *two* of these engineers.
3. The iron produced by Abraham Darby was too brittle and not suitable for the blacksmith. Henry Cort patented a process which overcame this problem and produced a malleable iron.
   Describe, with sketches, Cort's process.
4. Henry Bessemer discovered a method of making steel in large quantities which led to tremendous strides in engineering construction at the time of the Industrial Revolution.
   Describe, with the aid of sketches, the Bessemer process. Why did his method not prove popular in Britain?
5. Describe *two* engineering achievements of Joseph Bramah and the importance of his influence on the early work of Henry Maudslay.
6. *Either*
   (a) Sketch and describe the processes through which molten steel passes from the time it leaves the open-hearth furnace to its final rolling into steel sheet.
   Or
   (b) Describe the method of producing tool steel before the development of the Huntsman process.
7. (a) In 1709 Abraham Darby was responsible for one of the greatest steps forward in iron smelting history:
   (i) Write brief notes to explain how iron was smelted before 1709.
   (ii) List the major technical improvements in iron smelting in the two hundred years after 1709.
   (b) Draw a cross-section of an early eighteenth-century blast furnace.
8. Name and give details of five of Sir Joseph Whitworth's contributions to the field of production engineering.
9. *Either*
   (a) During the life of Henry Maudslay the standard and quality of metal cutting was greatly improved. Name *five* important contributions made in this field by Maudslay and, by brief notes, explain what the contributions were.
   Or
   (b) Draw a sketch to show how Henry Maudslay's first screw-cutting lathe (circa 1800) worked. List the advantages of this lathe over previous lathes.

(JMB O Level – Engineering Workshop Theory and Practice)

10. Explain briefly the method of operation of the steam engine developed by *one* of the following: Thomas Savery, Thomas Newcomen, James Watt. Compare the advantages, or deficiencies, of the engine you have chosen with the other two.
    Indicate briefly in what ways the steam engine influenced the development in this country of travel by land and sea?
11. Review briefly the position regarding iron and steel production in Britain during the fifteenth and sixteenth centuries and list the developments from then up to the end of the nineteenth century.
12. Choose a notable engineer concerned with the development of railways in Britain and write an essay on the importance of his work in this field.
13. Indicate briefly how water has been used as a source of energy from the thirteenth century to the present time to provide power for industrial purposes.
    Explain how coal has been used for similar purposes more recently and why it has tended to supersede water as a source of energy.
14. Choose an eminent engineer from the list given below and write an account of his work. Restrict your answer to the following headings:
    (a) list of items he invented or developed,
    (b) the effect of his work during his lifetime,
    (c) the influence of his work after his death.
    James Watt, Henry Maudslay, Sir Joseph Whitworth.
15. Trace briefly the historical development of the metal working lathe, mentioning in particular the following: pole lathe, mandrel lathe, slide rest, lead screw, back gearing, box-bed construction.
    Name, where applicable, the persons concerned with each item and the importance of each new development.
16. Describe *three* ways in which James Watt improved Newcomen's steam engine and explain the importance of each development.
    Steam engines have been used to power road vehicles, railway locomotives and factories. Explain why each application was supreme in its day but has now fallen generally into disuse.

(JMB A Level – Metalwork)

Name Index

Abbans, Marquis de, 75
Abercarn, tinning system, 32
Académie Royale, 103
Acheson, E. A., 45
Agricola, 96, 100
Ailnoth, 98
Alterneck, 87
Amalgamated Society of Engineers, 102-103
American System, the, 48
Ampere, 86
Anderson, Dr, 127
Arabia, as a source of technical knowledge, 98
Armstrong works, 121
Artificers, Statute of, 100
Arts, Society of, 107

Bedson, G., 28
Benier, 79
Benz, Karl, 82
Bessemer, Sir Henry, the converter, 15-16
  biography of, 112-113
Besson, 33
Beuth, P., 106
Birkbeck, G., 106
Black, Dr, 181
Bodmer, J., 43
Boulton, Matthew, 63, 66, 127, 128
Bramah, Joseph, 31, 46, 77
  biography of, 116-117
Branca, G., 70
Brown, Joseph, 43
  biography of the firm Brown & Sharpe, 123-125
Brunel, I. K., 104
  biography, 136-138
Brunel, Marc, 118
Buddle, J., 123
Burr, Thomas, 31

Calley, John, 60, 126
Carnegie, Andrew, 16
Carnot, N. L. S., 106
Carron foundry, 24, 127
Cartwright, the Rev. E., 74
Cayley, Sir George, 79
Charlottenburg, 106
China, as a source of technical knowledge, 98
Civil Engineers, Institution of, 104
Clark, Sir Dugald, 102
Clement, Joseph, 117, 118
Coalbrookdale, 4, 109
Combination Acts, 101
Cornish engines, 69
Cort, Henry, 7, 8
Cowles, 19
Cowper, E. A., 6
Crawshays, the 9
Creuzot works, 123
Curtis, C. G. 73

Daimler, Gottlieb, 82
Darby, Abraham I, 4, 24, 108, 109
  Abraham II, 4, 109
  Abraham III, 4, 109
  Edward William, 110
  biography of the Darby family, 108-110
Davenport, 87
Diesel, Rudolph, 85
'Dud' Dudley, 108
Durand, P., 31
Durer, Albrecht, 96

Ecole Polytechnique, 106
Egypt, and metals, 1
  and the bowdrill, 33
  and the lever, 49
  and measurement, 88
Electrical Engineers, Institute of, 105
Ellyot, William, 12
Engineering Employers Federation, 132
Eratosthones, 88
Ericsson, J., 79
Evans, Oliver, 68

Fairbairn, Sir William, 70
Faraday, Michael, 86
Farish, William, 106
Ferranti, S. Z., 20, 87
Fitch, Stephen, 40
Flanders, Walter, 139
Ford, Henry, biography of, 138-139
Ford, Richard, 109
Forest of Dean, 3
Fourneyron, B., 53
Fox, James, 38
Francis, J. B., 53
Freeth, J., 108
Fulton, Robert, 75

Galileo, 56
Galvani, L., 85
Gascoigne, William, 94
Gay and Silver, 42
Gilchrist, Percy, 115
Glehn, A. G. de, 75
Gramme, Z. T., 86
Great Eastern, the, 138
Great Western Railway, 137, 138
Greeks, the, and iron, 1
  and the lever, 47
  and water power, 50

Greeks, the *(continued)*
and measurement, 88
Guericke, Otto von, 57
Guerney, Sir G., 135

Hackworth, T., 75
Hall, J., 7
Hatfield, Sir Robert, 21
Heating and Ventilating Engineers, Institution of, 105
Hedley, William, 75
Henrici, hot-air engine, 80
Hero of Alexandria, 70
Héroult, P. L. T., 19
Hogge, Ralph, 23
Hornblower, Jonathan, 68, 129
Hornsby-Stuart, 84
Horsefall, J., 30
Howe, F. W., 42
Huntsman, Benjamin, 13
Huygens, C., 90

Iron Age, 1, 2

James of St George, 98
Johnson, Robert, 41
Jonval turbine, 53
Journeyman Steam Engine Makers, 101

Kelly, William, 15
Kjelin, 20

Lanchester, F. W., 83, 97
Laval, Gustave de, 73
Lavoisier, A. L., 106
Lee, Edmund, 55
Lenginour, Richard, 124
Lenoir, E., 80
Leupold, Jacob, 66
Levett, William, 23
Lincoln Miller, the, 42-43
Linz-Donawitz process, 19
Louth, Bernard, 28, 29
Lucas, Samuel, 9
Lunar Society of Birmingham, 103

Mackinley tariff, 32
Mannesman, the firm of, 31
Marine Engineers, 105
Martin, Pierre and Emile, 17, 114
Maudslay, Henry, and the lathe, 36-38
and measurement, 93, 95
biography of, 117-119
Maybach, W., 82
McNaught, John, 69
Mechanical Engineers, Institution of, 104
Mechanics' Institutes, 107
Melingriffith tinning system, 32
Meysey, Mathias, 12
Moissan, Henry, 19
Monge, Gaspard, 97, 106
Municipal Engineers, Institution of, 105
Murdock, William 66, 69, 80
Murray, Matthew, 47, 69
Mushet, Robert, 112

Nasmyth, James, and the steam hammer, 25, 26, 27
and the grinding machine, 45
and the milling machine, 41
and the safety ladle, 24
and the shaper, 47
biography of, 121-123
Neilson, J. B., 5
Newcomen, Thomas, his steam engine, 59-62
biography of, 125-126
New Stone Age, 1
Norse Mill, 50

Oersted, Hans, 85
Olds, Ransome, 138
Otto and Langen, 82
Otto, Nicholas, 81
Owen's College, Manchester, 107

Pacinotti, A., 86
Palmer, 124
Papin, Denis, 58, 60
Parsons, the Hon. Charles Algernon,
and the steam turbine, 71-73
biography of, 131-132
Pascal, B., 57, 77
Pelton Wheel, 53
Pen-y-darren, 66
Pickard, J., 65
Pickering, J., 25
Pixii, H., 86
Polheim, C., 8
Poncelot, T. V., 51
Portrush Railways, 87
Pratt, F. A., 42
Pratt and Whitney, 95
Priestman brothers, 84
Prussia and technical training, 105
Pyroscaphe, the, 75

Ramsden, Jesse, 36
Rankine, W. J. M., 107
Réaumur, R. A. F. de, 9
Réhé, 45
Reynolds, Richard, 109
Riley, James, 21
Ritchie, William, 107
Robbins and Lawrence, the firm of, 124
Roberts, Richard, 47, 118
Rochas, Alphonse Beau de, 81
Rocket, the, 135
Roebuck, John, 127
Romans, the, and metal, 12
and guilds, 98
and the crank, 50
and weights, 91
and mining, 98
Root, E. K., 42
Rupert, Prince, 9
Ruston and Hornsby, 84

Sa Hue Ra, 30

Savery, Thomas, and the steam engine, 58-59, 125, 126
Seguin, Marc, 75
Sellers, William, 95
Senot, Francois, 36
Sharpe, Lucien, 96
Siemens, Friedrich, 114
  Werner von, 87, 113
  Sir William, 16, 17, 18
  biography of, 113-115
Smeaton, John, 5, 52, 63
Sorel, M., 32
Southern, John, 66
Spencer, C., 40
Stephenson, George, biography of, 132-134
Stephenson, Robert, 78, 104
  biography of, 134-136
Stirling, Robert, 79, 114
Stockton and Darlington Railway, 134
Sun and Planet gear, 129
Symington, W., 75

Taylor, F. W., 21, 40
Telford, Thomas, 104
Thomas, Sydney Gilchrist, 18
  biography of, 115-116
Thomson, J., 54
Torricelli, E., 56
Trevithick, Richard, 66, 68, 70, 71
  biography of, 129-131
Turbinia, the, 132

Universal Miller, 43

Vaucanson, Jacques, 36, 41, 43
Vernier Calipers, 93-94
Vinci, Leonardo da, 33, 96
Vitruvius, 51, 96, 98
Volta, A., 85

Walter, R., 32
Watt, James, and the steam engine, 63-66
  attitude to high-pressure steam, 66
  biography of, 126-129
Weald of Sussex, 3
Wheaton, 45
Wheatstone, Sir Charles, 86
Whitney, Eli, 41, 48
Whitworth, Sir Joseph, and machine tools, 38, 47
  and measurement, 93, 94
  and screwthreads, 95
  and engineering education, 107
  biography of, 119-121
Wilkinson, David, toolmaker, 36
Wilkinson, Isaac, ironmaster, 109
Wilkinson, John, ironmaster, toolmaker, and industrialist, 8, 28, 30, 127, 128
  biography of, 110-111
Willis, Robert, 43, 106
Willmot, S. R., 124
Worcester, Marquis of, 59

Subject Index

Alloy steels, 21
Ancony, 4
Apprenticeship, 99, 100, 105, 106
Armourers, 98

Battery mill, 27
Blast furnace, 2-6
Boilers, 69-70
Boring machine, 44
Bow, the, as a tool, 33, 50
Bowl furnace, the, 1-2
Bronze, 1, 23

Casting iron, 8-9, 109
Casting metals, 23
Cementation, 12-13
Chafery, 4
Coating metals, 31, 32
Coke ovens, 4
Compound engines, 68-69
Condenser, used by Watt, 63
  used by Parsons, 72
Copper, the working of, 1
Crucible steel, 13-14
Cubit, 88
Cupola, the, 8

Diesel engine, 84
Drilling, 43
Drop forging, 25
D-type slide valves, 69

Electric furnaces, for steelmaking, 19-21
Electric power, 85-87
Engineering drawing, 96-97
Engineering, the profession, 98-107
Extrusion, 30-31, 117

Fantail, the, 55
Ferro-Manganese in steelmaking, 112
Finery, 3
First angle projection, 97
Forge, the, 3-4
Forging, 24-27
Four-stroke engines, 81

Gas engine, 79-80
Gold, 1
Grinding machines, 45-46
Guilds, 98-100

Hot-air engines, 79
Hydraulic power, 77-79

Impulse turbine, 73
Indicator, 66
Instrument makers, 92-93
Interchangeable parts, 116
Iron, its importance in the Industrial
  Revolution, 9-11
Ironmaking, by charcoal, 1-4
  by coke, 4-9

Journeymen, 100

Lathes, 33-40
Length, standards of, 88-91
Levers, 49
Linings, in steel furnaces, 48

Machine tools, 33-48
Machine tools – their economic importance, 48
Malleable cast iron, 9
Measurement, instruments of, 92-95
  standards of, 88-92
Metric System, 90
Mild steel, 15-16
Milling machines, 41-43
Millwrights, 100-101
Mining, 100

Oil engines, 81-82
Oliver hammers, 25
Open hearth furnace, the, 16-18
Orthographic projection, 97

Petrol engines, 82-84
Piecemaster system, 102
Pig iron, 6
Precision in engineering measurement, 93-96
Puddling furnace, 7-8

Reaction turbine, 71
Regeneration, 5, 17
Ringer, 3
Rolling, 28

Shaping machines, 46, 47
Shear steel, 13
Sheet metal and bars, 27
Silver, 1
Slitting of metal, 28
Standardisation in engineering, 94-96
Steam engines, 59-74
Steam engines, scientific background, 56-58
Steam hammer, 25, 123
Steam power, economic and social effects of, 73-76
Steam pump, the, 58-59
Steel, economic importance of, 22
Steel, production of, 11-22

Third angle projection, 97
Tilt hammer, 25
Tinplate, 31-32
Troy weight, 91
Tube drawing, 30-31
Turbines, steam, 70-73
Turbines, water, 53

Unions, 101–3

Water wheels, 50-53
Wedge, the, 49
Weight, standards of, 91-92
Wheel, the, 50
Windmills, 54-56
Wire drawing, 29-30
Wrought iron, 6, 8

Yard, the, 88–90

Iron and Steel

1709 DARBY I smelting with coke

1740 HUNTSMAN crucible furnace

Circa
1783 CORT's puddling process and rolling mill

1794 WILKINSON foundry cupola

1804 LUCAS malleable cast iron

1828 NEILSON hot blast

1839 NASMYTH's steam hammer

1856 BESSEMER converter
1857 COWPER stove patented

1862 BEDSON continuous rolling mill
1867 SIEMENS open-hearth process perfected
1878 GILCHRIST THOMAS basic furnace lining patented
1882 HATFIELD manganese steel

1889 RILEY nickel steel

1892 The improved Electric Arc Furnace

Machine Tools

1770 RAMSDEN's lathe with leadscrew
1774 WILKINSON's boring machine

Circa
1800 MAUDSLAY's improved screw-cutting lathe
1802 BRAMAH's wood-planer

1814 MURRAY's planer
Circa
1817 FOX's lathe
1817 ROBERT's lathe
Circa
1820 WHITNEY's miller
1820 CLEMENT's planer

1830's BODMER's gear cutters
1830 NASMYTH's miller

1834 NASMYTH's shaper
1835 WHITWORTH's lathe

1862 BROWN's universal miller
1876 BROWN's universal grinder

1891 ACHESON carborundum

Sources of Power

1663 MARQUIS OF WORCESTER's water-commanding engine

1698 SAVERY's steam pumping engine
1712 First NEWCOMEN engine erected
1745 The WINDMILL fantail
1769 WATT's condenser patented

1783 WATT's double acting rotary engine

1796 BRAMAH hydraulic press

1802 TREVITHICK's high pressure steam engine patented
1807 CAYLEY's hot air engine

1831 FARADAY demonstrates the principle of the dynamo

1845 MCNAUGHT's compound steam engine

1860 LENOIR gas engine
1861 DE ROCHES four stroke cycle

1876 OTTO gas engine

1884 PARSONS' steam turbine
1885 DAIMLER and BENZ petrol engine
1889 DE LAVAL's impulse turbine
1890 HORNSBY-STUART oil engine

1893 MAYBACH jet carburetter

1892/1895 DIESEL compression/ignition oil engine

Measurement

1790 Committee appointed which devised the METRIC system

1795 MONGE's book on descriptive geometry

1805 MAUDSLAY's micrometer

1841 WHITWORTH proposed adoption of standard screw thread in Britain

1851 BROWN & SHARPE vernier caliper

1856 WHITWORTH millionth-measuring machine

1868 BROWN & SHARPE pocket micrometer

1868 SELLERS standard screwthread in U.S.A.

1875 International Prototype Metre

1878 Imperial Standard Yard defined

1889 International Prototype Kilogramme

The Engineer and Craftsman

1794 ECOLE NORMALE Paris, first great institution for training engineers

1799–1800 THE COMBINATION ACTS suppressing unions in Britain

1818 Institution of Civil Engineers

1824 THE COMBINATION ACTS repealed

1826 Journeymen Steam Engine Makers Union

1846 Institution of Mechanical Engineers

1851 Amalgamated Society of Engineers

1872 Nine hour day in engineering
R.S.A. examinations

1882 Royal commission to examine state of technical education in Britain